PHYSICAL SCIENCE TEXTBOOKS
HISTORY AND PHILOSOPHY OF SCIENCE

PHYSICAL SCIENCE TEXTBOOKS
HISTORY AND PHILOSOPHY OF SCIENCE

MANSOOR NIAZ

Nova Science Publishers, Inc.
New York

Copyright © 2008 by Nova Science Publishers, Inc.

All rights reserved. No part of this book may be reproduced, stored in a retrieval system or transmitted in any form or by any means: electronic, electrostatic, magnetic, tape, mechanical photocopying, recording or otherwise without the written permission of the Publisher.

For permission to use material from this book please contact us:
Telephone 631-231-7269; Fax 631-231-8175
Web Site: http://www.novapublishers.com

NOTICE TO THE READER

The Publisher has taken reasonable care in the preparation of this book, but makes no expressed or implied warranty of any kind and assumes no responsibility for any errors or omissions. No liability is assumed for incidental or consequential damages in connection with or arising out of information contained in this book. The Publisher shall not be liable for any special, consequential, or exemplary damages resulting, in whole or in part, from the readers' use of, or reliance upon, this material.

Independent verification should be sought for any data, advice or recommendations contained in this book. In addition, no responsibility is assumed by the publisher for any injury and/or damage to persons or property arising from any methods, products, instructions, ideas or otherwise contained in this publication.

This publication is designed to provide accurate and authoritative information with regard to the subject matter covered herein. It is sold with the clear understanding that the Publisher is not engaged in rendering legal or any other professional services. If legal or any other expert assistance is required, the services of a competent person should be sought. FROM A DECLARATION OF PARTICIPANTS JOINTLY ADOPTED BY A COMMITTEE OF THE AMERICAN BAR ASSOCIATION AND A COMMITTEE OF PUBLISHERS.

LIBRARY OF CONGRESS CATALOGING-IN-PUBLICATION DATA

Niaz, Mansoor.
  Physical science textbooks : history and philosophy of science perspective / Mansoor Niaz.
    p. cm.
  ISBN 978-1-60456-485-3 (softcover)
  1. Physical sciences--History. 2. Physical sciences--Philosophy. 3. Physical sciences--Textbooks. I. Title.
  Q125.N487 2008
  500.209--dc22
                        2008004105

Published by Nova Science Publishers, Inc. ≃ *New York*

CONTENTS

Preface vii

Chapter 1 Introduction 1

Chapter 2 Nature of Science Framework
Based on History and Philosophy of Science 5

Chapter 3 Atomic Structure 9

Chapter 4 Determination of the Elemenary Electrical Charge 19

Chapter 5 Laws of Definite and Multiple Proportions
in Chemistry 25

Chapter 6 Kinetic Theory of Gases 29

Chapter 7 Van Der Waals' Equation of State:
a 'Progressive Problemshift' 33

Chapter 8 Kinetic Theory and Chemical Thermodynamics
as Rival Research Programs 35

Chapter 9 Understanding the Behavior of Gases:
from Hydrodynamic Laws to Kinetic Theory 37

Chapter 10 From 'Algorithmic Mode' to '
Conceptual Gestalt' in Students' Understanding
of the Behavior of Gases 39

Chapter 11 Criteria for Evaluation of Textbooks 41

Chapter 12 Covalent Bond 45

Chapter 13	Periodic Table	**51**
Chapter 14	Quantum Numbers	**63**
Chapter 15	Conclusion	**73**
References		**75**
Index		**87**

Preface

Research in science education has recognized the importance of presenting physical science within a history and philosophy of science (HPS) perspective in order to facilitate students' understanding. The objective of this study is to present a framework based on HPS for analyzing introductory freshman level general chemistry and physics textbooks published in U.S.A. A review of the relevant literature in science education shows that most textbooks ignore HPS, while presenting the following topics: a) Atomic structure; b) Determination of the elementary electrical charge; c) Laws of definite and multiple proportions; d) Kinetic theory of gases; e) Covalent bond; f) Periodic table; and g) Quantum numbers. This research shows that textbooks generally do not present progress in science as it is actually practiced by scientists. For most textbooks, doing science means accumulating empirical data with no reference to the interpretation of the data based on the scientist's theoretical framework or presuppositions. The role and importance of presuppositions and guiding assumptions along with the theory-ladenness of observations are the major contributions of the new post-positivist philosophy of science. Given the importance of textbooks in most parts of the world, such presentations deprive students of the dynamics of scientific research that involves controversies, conflicts and rivalries among scientists, that is in a nut shell the humanizing aspects of science. This research has been extended to show that inclusion of these facets of science in the classroom can be stimulating for students and facilitate greater conceptual understanding. It is concluded that if the authors want their students to understand science and not simply memorize algorithms, then a revision of the textbooks is necessary.

Chapter 1

INTRODUCTION

In an attempt to understand, What is Science?, The American Physical Society has drafted a policy statement, which has been endorsed by the American Association of Physics Teachers (AAPT, 1999): "Science is the systematic enterprise of gathering knowledge about the world and organizing and condensing that knowledge into testable laws and theories. The success and credibility of science is anchored in the willingness of scientists to abandon or modify accepted conclusions when confronted with more complete or reliable experimental evidence" (p. 659). For most practical purposes this definition is acceptable (most teachers at first glance may not find anything objectionable). Nevertheless, a closer look reveals the problematic nature of What is Science?, and hence the need for an epistemological framework based on history and philosophy of science (HPS). For example, the emphasis on 'gathering knowledge', i.e., experimental data and 'complete or reliable experimental evidence' with no reference to controversy, explanation of data by rival theories, and the tentative nature of scientific knowledge, shows the complexity of the issues involved. No wonder, one critic considers that by endorsing this statement the members of the Executive Board of the AAPT had, "... classified themselves as nonscientists" (Auerbach, 2000, p. 305). Furthermore, on the basis of an HPS perspective, this critic points out that the word 'complete' could be construed by Kuhnians (Kuhn, 1970) as, "... monotheistic Popperianism" (p. 305). AAPT is suggesting that progress in science is moving towards an increasing absolute truth. Among others, Kuhn (1970) and Lakatos (1970) have argued that scientific theories are not necessarily true or false, but rather are characterized by their heuristic power (ability to explain phenomenon).

AAPT statement has been a source of concern to science educators and another critic goes beyond by suggesting that we consult historians and philosophers of science on such issues, "... an important missing element in the definition is that it is perfectly rational and acceptable to 'believe' or accept as provisionally true, the best or most useful theory available. If we understand the history of science we can't help but think that better theories will emerge" (Forinash, 2000, p. 788).

Philosopher-physicist Gerald Holton (1996) has recommended the inclusion of history and philosophy of science in the science curriculum and raised a voice of concern with respect to the lack of interest in science courses. Philosopher-physical chemist Michael Polanyi (1964) has explained that although scientific discoveries are rich in controversy and interesting details, textbooks on the contrary fail to arouse students' interest:

> Yet as we pursue scientific discoveries through their consecutive publication on their way to the textbooks, which eventually assures their reception as part of established knowledge by successive generations of students, and through these by the general public, we observe that the intellectual passions aroused by them appear gradually toned down to a faint echo of their discoverer's first excitement at the moment of Illumination ... A transition takes place here from a heuristic act to the routine teaching and learning of its results, and eventually to the mere holding of these as known and true, in the course of which the personal participation of the knower is altogether transformed (pp. 171-172).

More recently, philosopher-physicist Stephen Brush (2000) has emphasized the importance of including history and philosophy of science in science textbooks. Similarly, research in science education has over the last many years, strongly endorsed the need for including history and philosophy of science in the curriculum (Burbules and Linn, 1991; Campanario, 2002; Clough, 2006; De Berg, 2003; Hodson, 1988; Irwin, 2000; Justi and Gilbert, 1999; Lederman et al. 2002; Matthews, 1994; Monk and Osborne, 1997; Niaz, 1994; Osborne et al. 2003).

Philosopher-physicist John Heilbron (1981a) has pointed out the difference in the recollections of a physicist and that of a historian, which clearly constitute an important guideline for the historical reconstruction of theories:

> The historian necessarily has a point of view different from that of a recollecting physicist. As Dirac discovered during a week spent with historians, we take an interest in precisely what the physicist wants (and manages) to forget, 'the various intermediate steps and ... false trails' ... The false trails taken

together lead more directly to the historian's goal of reconstructing the past state of science than the retrospectively clear highway of discovery (p. 223).

This precisely is the dilemma of science textbooks, as with retrospective hindsight all discoveries appear to be the work of geniuses who did not have to face the criticisms, rivalries, conflicts, passionate debates, and alternative interpretations of data by fellow scientists. Apparently, we have two options: leave the textbooks as they are, or to incorporate pertinent historical details (reconstructions) in order to present a picture of scientific progress based on human efforts and vicissitudes. This chapter follows the second alternative. Importance of textbooks and their evaluation has also been recognized by Project 2061's *Benchmarks for science literacy* (AAAS, 1993) and the *National standards for science education* (National Research Council, NRC, 1996). These documents call for the inclusion in textbooks of not only historical perspectives but also the introduction of terms meaningfully, appropriate representations of key ideas and the skilful use of models (Kesidou and Roseman, 2002).

The objectives of this chapter are: a) Present a framework based on history and philosophy of science (HPS) for analyzing introductory freshman general chemistry and physics textbooks published in U.S.A., and b) Review textbook studies based on the following topics: Atomic structure, Determination of the elementary electrical charge, Laws of definite and multiple proportions, Kinetic theory of gases, Covalent bond, Periodic table and Quantum numbers.

Chapter 2

NATURE OF SCIENCE FRAMEWORK BASED ON HISTORY AND PHILOSOPHY OF SCIENCE

Having recognized the importance of history and philosophy of science (HPS) in science education, considerable amount of research has been conducted in order to elaborate a framework for understanding nature of science (Abd-El-Khalick and Lederman, 2000; Alters, 1997; McComas et al. 1998; Niaz, 2001a; Scharmann and Smith, 2001; Smith et al. 1997; Smith and Scharmann, 1999; Solomon et al. 1996). A review of the literature shows that most teachers in many parts of the world lack an adequate understanding of nature of science (Akerson et al. 2006; Bell et al. 2001; Blanco and Niaz, 1997; Clough, 2006; Lederman, 1992; Mellado et al. 2006; Pomeroy, 1993). This should be no surprise to anyone who has analyzed science curricula and textbooks, which are heavily stanced towards an entirely empirical and positivist epistemology. Many science teachers would perhaps argue that nature of science has no place in science courses or textbooks. Marquit (1978) has exposed this fallacy in eloquent terms, "... textboks in physics invariably have a philosophical content whether or not the authors are conscious of it, ... In this way students are taught possibly highly controversial philosophical viewpoints [empirical nature of science] with no indication that there are alternative outlooks" (p. 784).

Schwab (1962, 1974) differentiates between the methodological (empirical data) and interpretative (heuristic principles) components of scientific knowledge. It is not verified knowledge (accumulation of empirical data) but rather the heuristic principles that help us to structure inquiry. Schwab's idea of a heuristic principle (construction of the mind) comes quite close to what contemporary philosophers of science have referred to as presuppositions (Holton, 1978), hard core (Lakatos, 1970), or guiding assumptions (Laudan et al.1988). In the case of the physical sciences, Schwab (1974) has specifically elaborated with respect to the significance of heuristic principles, "In physics, similarly, we did not know from the beginning that the properties of matter are fundamental and determine the behavior of these particles; their relations to one another. It was not verified knowledge but a heuristic principle, needed to structure inquiry, that led us to investigate mass and charge and, later, spin" (p. 165).

Let us understand Schwab's idea of a heuristic principle within the context of a particular historical episode. When J.J. Thomson, a British physicist, started his work on cathode rays he was fully aware of the controversy with regards to the nature of cathode rays: were they particles or waves in the ether (Falconer, 1987). Thomson (1897) in his now famous article expressed the dilemma en lucid terms:

> As the cathode rays carry a charge of negative electricity, are deflected by an electrostatic force as if they were negatively electrified, and are acted on by a magnetic force in just the way in which this force would act on a negatively electrified body moving along the path of these rays, I can see no escape from the conclusion that they are charges of negative electricity carried by particles of matter. The question next arises, What are these particles? Are they atoms, or molecules, or matter in a still finer state of subdivision? To throw some light on this point, I have made a series of measurements of the ratio of mass of these particles to the charge carried by it. To determine this quantity, I have used two independent methods (p. 302).

This was the crucial aspect of Thomson's article and he clearly visualized that the determination of the mass (m) to charge (e) ratio (m/e) of the cathode rays would help him to identify them as ions or a universal charged particle. This precisely constitutes a heuristic principle for Schwab, viz., what guides a scientist to do experiments and collect data. Thomson's peers questioned his interpretations and not the data, which led to considerable controversy. Textbooks generally ignore such principles and simply emphasize the experimental details, with no reference to the difficulties faced by the scientist. On the contrary, Thomson's (1897) article in the *Philosophical Magazine*, can indeed be considered a masterpiece of scientific and pedagogical reasoning and one can see how difficult it is to interpret experimental data.

Interestingly, two German physicists, Kaufmann (1897) and Wiechert (1897) also determined the (m/e) ratio of cathode rays in the same year as Thomson and their interpretations agreed with each other. If we tell students that 'science is empirical' (this is what most textbooks do), we shall be denying students an important aspect of the nature of science, viz., what made Thomson's work different from that of Kaufmann and Wiechert. Falconer (1987) has explained the difference cogently:

> Kaufmann, an ether theorist, was unable to make anything of his results. Wiechert, while realizing that cathode ray particles were extremely small and universal, lacked Thomson's tendency to speculation. He could not make the bold, unsubstantiated leap, to the idea that particles were constituents of atoms. Thus, while his work might have resolved the cathode ray controversy, he did not 'discover the electron' (p. 251).

I have gone into considerable historical detail in order to illustrate the empirical nature of science. At this stage I will present a consensus view of the nature of science within the science education community. As the historical episodes unfold, I will illustrate the different aspects of the nature of science. Despite some controversy, nature of science can be characterized by the following aspects (Niaz, 2007):

1. Empirical nature of science: Scientific knowledge relies on observations and experimental evidence. However, interpretations of the data based on heuristic principles, rational arguments and skepticism are equally important.
2. Observations are theory-laden.
3. Science is tentative/fallible.
4. There is no one way to do science and hence no universal step-by-step scientific method can be followed.
5. Laws and theories serve different roles in science and hence theories do not become laws even with additional evidence.
6. Scientific progress is characterized by competition among rival theories.
7. Different scientists can interpret the same experimental data in more than one way.
8. Development of scientific theories at times is based on inconsistent foundations.
9. Scientists require accurate record keeping, peer review and replicability.
10. Scientists are creative and often resort to imagination and speculation.
11. Scientific ideas are affected by their social and historical milieu.

Chapter 3

ATOMIC STRUCTURE

The history of structure of the atom since the late 19th and early 20th century shows that the atomic models of J.J. Thomson, E. Rutherford, and N. Bohr evolved in quick succession and had to contend with controversies based on rival research programs. At a Symposium 'History of the atom' held at the annual meeting of the American Physical Society and the AAPT, Chicago, January 1980, Gerald Holton emphasized the importance of historical reconstruction in order to understand our present theories of atomic structure, "... recount the intellectual feats and defeats over more than 2500 years, from Thales to Bohr, without which we would not have gained the current bright prospects" (Holton, 1981, p. 25).

THOMSON'S MODEL OF THE ATOM

In order to understand Thomson's model of the atom textbooks need to refer to the following heuristic principles (criteria):

Cathode Rays as Charged Particles or Waves in the Ether

Thomson's experiments were conducted against the backdrop of a conflicting framework. Thomson (1897) explicitly points out that his experiments were conducted to clarify the controversy with regard to the nature of cathode rays, that is, charged particles or waves in the ether. Resolution of this controversy constituted a heuristic principle for Thomson's research program and its inclusion in the textbooks could help students to understand as to why scientists do experiments. Niaz (1998) has shown that of the 23 general chemistry textbooks analyzed only two (Bodner and Pardue, 1989; Oxtoby et al. 1990) made a simple mention to this controversy. Rodríguez and Niaz (2004a) have shown that of the 41 general physics textbooks analyzed only two (Cohen, 1976; Ohanian, 1987) simply mentioned the controversy and two (Cooper, 1970; Jones and Childers, 1990) made a satisfactory presentation.

Determination of Mass-to-Charge Ratio to Decide Whether Cathode Rays were Ions or Universal Charged Particles

The heuristic principle that helped Thomson to postulate his model of the atom, dealt with as to how and why he determined the mass (m) to charge (e) ratio (m/e) of cathode rays. Niaz (1998) has reported that of the 23 general chemistry textbooks only two described satisfactorily that Thomson determined the ratio (m/e) to decide whether cathode rays were ions or a universal charged particle. Following is an example of a satisfactory presentation:

> A very striking and important observation made by Thomson is that the e/m ratio does not depend on the gas inside the tube or the metal used for the cathode or anode. The fact that the e/m ratio is the same whatever gas is present in the tube proves that the cathode ray does not consist of gaseous ions, for it did, e/m would depend on the nature of the gas (Segal, 1989, p. 412).

Let us now contrast this with a textbook that made no mention of the heuristic principle:

> A physicist in England named J.J. Thomson showed in the late 1890s that the atoms of any element can be made to emit tiny negative particles. (He knew they had a negative charge because he could show that they were repelled by the negative part of an electric field). Thus he concluded that all types of atoms must contain these negative particles, which are now called *electrons* (Zumdahl, 1990, p. 97).

This presentation makes no attempt to present arguments, reasons or strategies used by Thomson in order to postulate his model of the atom, and Schwab (1962) very aptly refers to as a 'rhetoric of conclusions.'

Rodríguez and Niaz (2004a) have shown that of the 41 general physics textbooks evaluated only one (Eisberg, 1973) made a simple mention and one (Cooper, 1970) made a satisfactory presentation. Cooper (1970) presents a detailed reconstruction that follows the sequence: Historical context --- Thomson's understanding of the controversy --- Experimental details --- Thomson's strategy to tackle the dilemma --- Mathematical details --- Postulation of the electron. Finally, Cooper (1970) summarizes his reconstruction in the following terms: "Not the first time that the greatest difficulty in doing a crucial experiment lay as much in developing necessary techniques as in its conception" (p. 312), which can provide guidelines for future textbooks.

RUTHERFORD'S MODEL OF THE ATOM

In the very first paragraph of his famous article in the *Philosophical Magazine* Rutherford (1911) starts on controversial note: "It has generally been supposed that the scattering of a pencil of alpha or beta rays in passing through a thin plate of matter is the result of a multitude of small scatterings by the atoms of matter traversed" (p. 669). This of course referred to the experimental work of Crowther (1910), a colleague of Thomson. Rutherford (1911) explicitly points out that Crowther's experimental results provided support for Thomson's hypothesis of compound scattering. This served as a preamble for Rutherford (1911) to present his side of the story:

> The observations, however, of Geiger and Marsden (1909) on the scattering of alpha rays indicate that some of the alpha particles must suffer a deflexion of more than a right angle at a single encounter ... that a small fraction of the incident alpha particles, about 1 in 20,000 were turned through an average angle of 90° in passing through a layer of gold-foil ... A simple calculation based on the theory of probability shows that the chance of an alpha particle being deflected through 90° is vanishingly small ... A simple calculation shows that *the atom must be a seat of an intense electric field in order to produce such a large deflexion at a single encounter* (p. 669, emphasis added)

Rutherford had the experimental data as early as June 1909, to postulate his model of the nuclear atom, and yet he did not do so until March 1911. What happened between June 1909 and March 1911 is important not only for historians and philosophers of science, but also for science teachers. Soon after Geiger and Marsden (1909) published their results, Thomson and colleagues also started working on the scattering of alpha particles in their laboratory. Although experimental data from both laboratories were similar, interpretations of Thomson and Rutherford were entirely different. Thomson propounded the hypothesis of *compound scattering*, according to which a large angle deflection of an alpha particle resulted from successive collisions between the alpha particles and the positive charges distributed throughout the atom. Rutherford in contrast, propounded the hypothesis of single *scattering*, according to which a large angle deflection resulted from a single collision between the alpha particle and the massive positive charge in the nucleus. The rivalry led to a bitter dispute between the proponents of the two hypotheses. Rutherford even charged Crowther (1910), a colleague of Thomson, to have 'fudged' the data in order to provide support for Thomson's model of the atom (Wilson, 1983). Heilbron (1981b) has provided a succinct account of the rivalry between Thomson and Rutherford:

> Now in 1910 Thomson was the undisputed world master in the design of atoms, and neither Bragg nor Rutherford had yet tried their hand at the type of quantitative model making at which he excelled: they were challenging the acknowledged leader of English physics, their own former teacher, in what had been his most exclusive preserve. And Rutherford was coming into open conflict with Thomson's ideas for the first time. The situation thus did not lack a competitive aspect, and this, I think helps explain the harshness of the attacks against Crowther (p. 134).

Rutherford's dilemma: On the one hand he was entirely convinced and optimistic that his model of the atom explained experimental findings better, and yet it seems that the prestige, authority, and even perhaps some reverence for his teacher made him waver. However, in a letter to Schuster (Secretary of the Royal Society), written about 3 years later (February 2, 1914), Rutherford is much more forceful:

> I have promulgated views on which J.J. [Thomson] is, or pretends to be, skeptical. At the same time I think that if he had not put forward a theoretical atom himself, he would have come round long ago, for the evidence is very strongly against him. If he has a proper scientific spirit I do not see why he should hold aloof and the mere fact that he was in opposition would liven up the meeting (Reproduced in Wilson, 1983, p. 338).

This makes interesting reading. A science student may wonder as to why Thomson and Rutherford did not meet over dinner (they were well known to each other) and decide in favor of one or the other model. Progress in science is, however, much more complex. Both, Thomson and Rutherford, stuck to their presuppositions. Again, a student may wonder as to how Rutherford could doubt the 'proper scientific spirit' of none else but the world master in the design of atomic models. These issues, if discussed in class and textbooks, could make science much more human and attractive. In order to understand Rutherford's model of the atom, textbooks need to refer to the following heuristic principles:

Probability of Large Angle Deflections is Exceedingly Small as the Atom is the Seat of an Intense Electric Field

The crucial argument that clinched the debate in favor of Rutherford's model was not the large angle deflection of alpha particles, but rather the knowledge that 1 in 20,000 particles deflected through large angles. Niaz (1998) has shown that of the 23 general chemistry textbooks analyzed only 2 described this heuristic principle satisfactorily and the following is an example:

....Using a gold foil 0.00004 cm thick, he found that one alpha particle in 20,000 was deflected through an angle greater than 90°. From such experiments, Rutherford concluded that since most of the alpha particles pass through the foil undeflected, the volume occupied by an atom must consist largely of empty space (Sisler et al. 1980, p. 164).

Rodríguez and Niaz (2004a) have reported that of the 41 general physics textbooks analyzed, 5 described satisfactorily and 9 made a simple mention.

Single/Compound Scattering of Alpha Particles

To maintain his model of the atom and to explain large angle deflections of alpha particles, Thomson put forward the hypothesis of compound scattering. Niaz (1998) has shown that of the 23 general chemistry textbooks analyzed none described satisfactorily or mentioned this principle. Rodríguez and Niaz (2004a) have reported that of the 41 general physics textbooks analyzed, 2 described satisfactorily and 2 made a simple mention. Cooper (1970) presented a detailed satisfactory presentation:

Rutherford calculated that from the large Thomson positive charge distribution particles should never be deflected more than 0.03 degrees in a single collision; in undergoing multiple collisions they should have about an equal chance of being deflected one way as another. Therefore, large deflections as a result of many single deflections in the same direction were very improbable. (It had been calculated on the basis of the Thomson model that a total deflection greater than 90° in traversing the gold foil would have only one chance in 10^{3500} of occurring) (p.321).

Bohr's Model of the Atom

On the very first page of his now famous trilogy published in the *Philosophical Magazine*, Bohr (1913) starts on a controversial note by pointing out the difficulties associated with Rutherford's model of the atom:

> In an attempt to explain some of the properties of matter on the basis of this atom-model [Rutherford's] we meet, however, with difficulties of a serious nature arising from the apparent instability of the system of electrons: difficulties purposely avoided in atom-models previously considered, for instance, in the one proposed by Sir J.J. Thomson (p. 2).

In the fourth paragraph Bohr (1913) formulates his epoch-making postulate:

> The result of the discussion of these questions seems to be a general acknowledgment of the inadequacy of the classical electrodynamics in describing the behaviour of systems of atomic size ... it seems necessary to introduce in the laws in question a quantity foreign to the classical electrodynamics, i.e., Planck's constant, or as it often is called the *elementary quantum of action* (p. 2, emphasis added).

Contrary to textbook accounts, according to Lakatos (1970):

> Bohr's problem was not to explain Balmer's and Paschen's series, but to explain the paradoxical stability of the Rutherford atom. Moreover, Bohr had not even heard of these formulae before he wrote the first version of his paper (p. 147).

The 'first version' Lakatos is referring to is of course the 'Rutherford Memorandum' written by Bohr for Rutherford in June-July of 1912, and is considered to be a crucial document by Heilbron and Kuhn (1969). A letter, written by Bohr to Rutherford on Jan. 31, 1913, shows that even then he was not fully aware of the implications of spectroscopic research for his problem:

> I do not at all deal with the question of calculation of the frequencies corresponding to the lines in the visible spectrum. I have only tried, on the basis of the simple hypothesis, which I used from the beginning, to discuss the

constitution of the atoms and molecules in their 'permanent state' [Reproduced by Rosenfeld, 1963, pp. xxxvi – xxxvii].

A reconstruction of the events that led Bohr to postulate his model of the atom has important implications for our understanding of scientific progress and practice. According to Lakatos (1970):

> Since the Balmer and the Paschen series were known before 1913 [year of Bohr's first publication], some historians present the story as an example of a Baconian 'inductive ascent':
>
> 1) the chaos of spectrum lines,
> 2) an 'empirical law' (Balmer),
> 3) the theoretical explanation (Bohr) (p. 147).

The inductive ascent refers to the positivist/inductivist interpretation of the history of science, according to which scientific progress follows the sequence: experimental observations, scientific laws and theories.

Another important aspect of Bohr's model of the atom is the presence of a deep philosophical chasm: that is, in the stationary states, the atom obeys classical laws of Newtonian mechanics; on the other hand, when the atom emits radiation, it exhibits discontinuous (quantum) behavior. Based on these and other arguments, Bohr's 1913 article, in general, had a fairly adverse reception in the scientific community. Rutherford, although no philosopher of science, was the first to point this out, when he wrote to Bohr on March 20, 1913:

> The mixture of Planck's ideas with the old mechanics makes it very difficult to form a physical idea of what is the basis of it all ... How does the electron decide what frequency it is going to vibrate at when it passes from one stationary state to another? (Reproduced in Holton, 1993, p. 80).

Otto Stern told a friend: "If that nonsense is correct that Bohr has just published, then I will give up being a physicist" (Holton, 1986, p. 145). H.A. Lorentz objected: "... the individual existence of quanta in the aether is impossible" (Holton, 1993, p. 79).

Lakatos (1970) has argued that Bohr employed a methodology used frequently by scientists in the past and perfectly valid for the advancement of science:

> *Some of the most important research programmes in the history of science were grafted on to older programmes with which they were blatantly inconsistent*. For instance, Copernican astronomy was 'grafted' on to Aristotelian physics, Bohr's programme on to Maxwell's. Such 'grafts' are irrational for the justificationist and for the naive falsificationist, neither of whom can countenance growth on inconsistent foundations... As the young grafted programme strengthens, the peaceful co-existence comes to an end, the symbiosis becomes competitive and the champions of the new programme try to replace the old programme altogether (p. 142, original italics).

In order to understand Bohr's model of the atom, textbooks need to refer to the following heuristic principles:

Paradoxical Stability of the Rutherford Model of the Atom

Bohr's main objective was to explain the paradoxical stability of the Rutherford model of the atom, which constituted a rival framework for his own model. Niaz (1998) has shown that of the 23 general chemistry textbooks analyzed, 4 mentioned that Bohr's main objective was to explain the paradoxical stability of the Rutherford model of the atom, which constituted a rival framework and 3 described it satisfactorily. Following is an example of a satisfactory presentation:

> ... Description of the atom, which is universally accepted today, seemed very surprising and unlikely to the scientists of 1911. Why should such a structure be stable? Positive and negative charges attract one another --- what keeps the negative electrons at some distance from the positive nucleus? Why aren't the electrons drawn into the nucleus as a result of the coulombic (electrostatic) attraction? Indeed, the laws of classical electromagnetic theory ... predict unequivocally that an atom with such a structure could not exist. Nothing in classical physics can explain the existence of a stable atom with the structure elucidated by Rutherford (Segal, 1989, p. 415).

Rodríguez and Niaz (2004a) have reported that of the 41 general physics textbooks analyzed only one (Cooper, 1970) explained satisfactorily that Bohr's main objective was to explain the paradoxical stability of the Rutherford model of the atom. Cooper (1970) presents a thought-provoking reconstruction of the arguments that led Bohr to introduce the 'quantum of action' and concludes in the following terms:

The requirement of Maxwell's theory that an electron accelerating about a positive central charge should radiate energy undermined the possibility for the stability of the Rutherford atom from the point of view of classical physics ... This collection of facts, interpreted via existing theory, seemed to lead to a dead end at every turn... in retrospect ... Rutherford's atom was the ultimate application of purely classical principles in the atomic domain. As though he had opened the seventh seal, there was a silence ... and the angels were given seven trumpets ... (p. 324).

Explanation of the Hydrogen Line Spectrum

Bohr had not even heard of the Balmer and Paschen formulae for the hydrogen line spectrum, when he wrote the first version of his 1913 article. Failure to understand this episode within a historical perspective led to an inductivist/positivist interpretation, referred to as the 'Baconian inductive ascent' by Lakatos (1970). Niaz (1998) has shown that of the 23 general chemistry textbooks analyzed none described satisfactorily the quantization of the Rutherford model of the atom by Bohr. Rodríguez and Niaz (2004a) have reported that none of the general physics textbooks described satisfactorily that Bohr had not even heard of the Balmer and Paschen formulas for the hydrogen line spectrum.

Deep Philosophical Chasm in Bohr's Model

Bohr's incorporation of Planck's 'quantum of action' to the classical electrodynamics of Maxwell, represented a strange 'mixture' for many of Bohr's contemporaries and philosophers of science. This illustrates how scientists when faced with difficulties, often resort to such contradictory 'grafts'. Niaz (998) has shown that of the 23 general chemistry textbooks analyzed, four simply mentioned this heuristic principle and two made a satisfactory presentation, of which following is an example:

> There are two ways of proposing a new theory in science, and Bohr's work illustrates the less obvious one. One way is to amass such an amount of data that the new theory becomes obvious and self-evident to any observer. The theory then is almost a summary of the data. The other way is to make a bold new assertion that initially does not seem to follow from the data, and then to demonstrate that the consequences of this assertion, when worked out, explain

many observations. With this method, a theorist says, "You may not see why, yet, but please suspend judgment on my hypothesis until I show you what I can do with it." Bohr's theory is of this type. Bohr said to classical physicists: "You have been misled by your physics to expect that the electron would radiate energy and spiral into the nucleus. Let us assume that it does not, and see if we can account for more observations than by assuming that it does" (Dickerson, et al., 1984, p. 264).

Rodríguez and Niaz (2004a) have reported that of the 41 general physics textbooks analyzed, only one described satisfactorily that Bohr's incorporation of Planck's 'quantum of action' was based on an inconsistent foundation and represented a 'deep philosophical chasm'. Cooper (1970) provides an example of a satisfactory presentation:

> Focusing his attention entirely on the construction of a nuclear atom, Bohr took what principles of classical physics he needed and added several nonclassical hypotheses almost without precedent; the mélange was not consistent. But they formed a remarkably successful theory of the hydrogen atom. It would be years before it could be said that one had a consistent theory again (p. 325).

The importance of including history and philosophy of science (HPS) in science education, was recognized as early as 1932, in an editorial in the *Journal of Chemical Education*, published by the American Chemical Society (Reinmuth, 1932). Thus, we decided to analyze general chemistry textbooks published between 1929-1967, based on the same heuristic principles as our studies based on more recently published textbooks. Rodríguez and Niaz (2002) analyzed 30 general chemistry textbooks published between 1929-1967 and found that these not only ignore HPS, but also present experimental findings as a 'rhetoric of conclusions' (Schwab, 1962), quite similar to more recent textbooks.

Results reported in this section show that in the presentation of the atomic models of Thomson, Rutherford and Bohr, general chemistry and physics textbooks systematically present an inductivist perspective. A few of the textbooks, without following any particular historical approach, present atomic models in a context that approximates to recent interpretations of HPS. The textbook by Cooper (1970) in this respect is quite remarkable and can provide guidelines for future textbooks.

Chapter 4

DETERMINATION OF THE ELEMENARY ELECTRICAL CHARGE

A historical reconstruction of the oil drop experiment that led to the determination of the elementary electrical charge (e) shows the controversial nature of the experiment then (1910-1925) and that the experiment is difficult to perform even today (Jones, 1995). After almost 90 years, historians and philosophers of science do not seem to agree what really happened (Niaz, 2005). Despite these difficulties, most chemistry and physics textbooks consider the oil drop experiment to be a simple, classic and a beautiful experiment, in which Robert A. Millikan (1868-1953) by an exact experimental technique determined the elementary electrical charge, viz., charge of the electron.

An important aspect of Millikan's experiments (University of Chicago) is that he clearly formulated the guiding assumptions (hard core, Lakatos, 1970) of his research program, from the very beginning. Millikan recounts how after reading the publications of Townsend (1897) and Thomson (1898) he became interested in the determination of the elementary electrical charge and at the turn of the century formulated the following research questions (Millikan, 1947, p. 41): a) What are the masses of the constituents of the atoms torn asunder by X-rays and similar agencies? b) What are the values of the charges carried by these constituents; c) How many of these constituents are there? d) What volumes do they occupy? e) Do all atoms possess similar constituents? In other words, is there a primordial sub-atom out of which atoms are made? The last question, of course referred to J.J. Thomson's finding that the mass to charge (m/e) ratio is independent of the gas in the cathode ray tube. This precisely set the stage for Millikan's determination of the elementary electrical charge (e). He outlined his research problem in terms that can easily be interpreted as his guiding assumption:

"... whether the electron which had first made its appearance in Faraday's experiments on solutions and then in Townsend's and Thomson's experiments on gases is after all only a *statistical mean* of charges which are themselves divergent" (Millikan, 1947, p. 58). This, of course, implied the existence of the elementary electrical charge, and this is what Millikan pursued tenaciously for almost 25 years.

Felix Ehrenhaft (1879-1952), Millikan's major critic did most of his experimental work at the University of Vienna and was considered in 1910 (when the controversy started) a fairly well established figure in the European scientific community. In Ehrenhaft's (1910) first major criticism he closely scrutinized Millikan's (1910) data. He recalculated the charge on each drop from each of Millikan's observations separately. Millikan (1910), in contrast, used average values of times of ascent and descent, measured on different droplets. Ehrenhaft's calculations produced a large spread of values of the elementary electrical charge, ranging from 8.60×10^{-10} esu to 29.82×10^{-10} esu. Furthermore, Ehrenhaft showed how Millikan's method led to paradoxical situations. A drop with a charge of $e_n = 15.59 \times 10^{-10}$ esu had been placed among those assumed to be carrying three electrons ($n = 3$), whereas another drop with a charge of $e_n = 15.33 \times 10^{-10}$ esu was assumed to be carrying four electrons ($n = 4$). How do we explain these differences?

Holton (1978) provides the following insight on the impasse: "It appeared that the same observational record could be used to demonstrate the *plausibility of two diametrically opposite theories,* held with great conviction by two well equipped proponents and their respective collaborators" (pp. 199-200, emphasis added). Ehrenhaft wrote about a dozen articles in the following 4 years, all implicitly aimed at discrediting Millikan's measurements. Millikan also wrote extensively and rebutted Ehrenhaft's criticisms. In order to have a flavor of the arguments presented by Millikan (1916) in his defense, the following is an example:

> That these same ions have one sort of charge when captured by a big drop and another sort when captured by a little drop is obviously absurd. If there are not the same ions which are caught in the two cases, then, *in order to reconcile the results with the existence of the exact multiple relationship ..., it would be necessary to assume that there exist in the air an infinite number of different kinds of ionic charges corresponding to the infinite number of possible radii of drops, and that, when a powerful electric field drives all of these ions towards a given drop, this drop selects in each instance just the charge which corresponds to its particular radius.* Such an assumption is not only too grotesque for serious

consideration but is directly contradicted by my experiments ... (p. 617, original italics).

This passage is indeed revealing. Millikan is telling the reader that experimental observations are important, but there is something even more important, viz., the guiding assumptions, and any data that go against them would appear to be 'absurd' and 'grotesque' and hence subelectrons (Ehrenhaft's thesis) could not exist. Millikan's data showed that all drops had a charge that was an integral multiple of the elementary electrical charge, thus providing evidence for the existence of the electron. Research literature at the time considered Rutherford and Geiger's value of $e = 4.657 \times 10^{-10}$ esu as the most probable value. Millikan's data provided a value of e very close to this. Ehrenhaft, too, obtained data that he interpreted as integral multiple of the elementary electrical charge (e). Nevertheless, his argument was precisely that there were many drops that did not lead to an integral multiple of e.

Millikan's Hand-Written Notebooks and the Controversy

A new dimension to the controversy was added by Holton's (1978) discovery of Millikan's two laboratory notebooks in his Archives at the California Institute of Technology, Pasadena. These notebooks have raw data and some of the data reduction procedures used in his *Physical Review* article (Millikan, 1913). The Millikan notebooks (October 28, 1911 to April 16, 1912, about 175 pages) are indeed a rare opportunity to see a scientist working in his laboratory. Holton (1978, p. 207) reproduced the data in Millikan's handwriting from one of the 140 experiments that are included in the notebooks. Apparently, this was an experiment that did not give value of e that Millikan was expecting, and he noted frankly: "*Error high* will not use ... can work this up and probably is ok but point is [?] not important. Will work if have time Aug. 22." Holton (1978) remarked on this experiment: "It was a failed run ---, *or effectively, no run at all*. Instead of wasting time investigating it further, he simply went on to make another set of readings with a new drop, recorded on the next page of the notebook"(p. 209).

Now let us turn to the actual publication, in which Millikan (1913) meticulously presented complete data on 58 drops and emphasized that all of the drops experimented upon had been included (p. 138). The laboratory notebooks tell us that there were 140 drops and the published results are emphatic that there were 58 drops. What happened to the other 82 drops? Herein lies the crux of the difference between the methodologies of Ehrenhaft and Millikan. What was the warrant under which Millikan discarded more than half of his observations? The answer is simple but not found frequently in the research literature, and much less in textbooks. Millikan's guiding assumptions provided the warrant. Indeed, Millikan would perhaps have liked to warn Ehrenhaft that all the readings cannot be used as their experiments were constantly faced with difficulties such as evaporation, sphericity, radius, and change in density of drops due to oxides or dust particles, and variation in experimental conditions (battery voltages, stopwatch errors, temperature, pressure, and convection). In order to understand Millikan's determination of the elementary electrical charge, textbooks need to refer to the following heuristic principles.

Millikan-Ehrenhaft Controversy

Niaz (2000a) has shown that of the 31 general chemistry textbooks analyzed none mentioned the controversy. Similarly, Rodríguez and Niaz (2004b) have reported that of the 43 general physics textbooks analyzed none mentioned the controversy. Some textbooks explicitly denied that the drops studied by Millikan had fractional charges, i.e., a charge unequal to an integer times the electron charge and following is an example: "Millikan found that each droplet had a charge of either zero, e, $2e$, $3e$, $4e$, or some whole number times e, but never a fraction of e...No one has ever found an electric charge that is not a whole number times the elementary charge, the charge on the electron" (Hulsizer and Lazarus, 1977, p. 227).

Millikan's Guiding Assumption

Based on the atomic nature of electricity, Millikan hypothesized the existence of an elementary electrical charge. Millikan found drops with a wide range of electrical charges. Despite such anomalous data, if it were not for the guiding assumption, Millikan would have abandoned the search for the elementary electrical charge. Niaz (2000a) has shown that of the 31 general chemistry textbooks analyzed only 6 made a simple mention. Rodríguez and Niaz (2004b) have reported that of the 43 general physics textbooks analyzed only 2 made a simple mention and following is an example: "By observing the motion of the hundreds of droplets with different charges on them, Millikan *uncovered the pattern he expected*: the charges were multiples of the smallest charge he measured" (Olenick, et al., 1985, p. 241, emphasis added). 'The pattern' can be considered as an oblique reference to his guiding assumption. The same textbook reproduced the following quote from Millikan's laboratory notebook: "One of the best ever [data] ... almost exactly *right*. Beauty – publish." After reproducing the quote, authors asked a very thought provoking question: "What's going on here? How can it be right if he's supposed to be measuring something he doesn't *know*? One might expect him to publish everything!" (p. 244). This is the closest that any textbook came to with respect to mentioning Millikan's guiding assumption.

Dependence of the Elementary Electrical Charge on Experimental Variables

Millikan was constantly trying to improve his experimental conditions to obtain the charge on the droplets as an integral multiple of the elementary electrical charge. Some of the variables that he constantly referred to were evaporation, sphericity, and radius of the droplets, change in density of the droplets, changes in battery voltages, temperature, and viscosity of the air. Niaz (2000a) has shown that of the 31 general chemistry textbooks analyzed only 2 presented this heuristic principle satisfactorily and following is an example:

> Using an atomizer, microscopic spherical drops of oil are introduced into the space above two charged plates. Oil is used because it does not noticeably evaporate ... Gravity causes the drops to fall, but they are slowed by friction, due to the viscosity of the air ... The charge on the drop can be calculated from the value of its downward velocity, the magnitude of the potential difference, the known acceleration of gravity, the density of oil, and the air viscosity (Segal, 1989, pp. 412-413).

Without the description of these and other experimental variables that made the oil drop experiment so difficult to perform, students are left with the impression that in order to discover new things, scientists need only to walk into the laboratory. Rodríguez and Niaz (2004b) have reported that only one textbook made a mention.

Millikan's Experiments as Part of a Progressive Sequence of Heuristic Principles

Millikan's work started by repeating and a critical evaluation of the experimental work of Townsend, Thomson, and Wilson on charge clouds of water droplets. Each stage in this historical process was characterized by guiding assumptions, improvement in experimental techniques, criticisms and rebuttals. Niaz (2000a) has shown that of the 31 general chemistry textbooks analyzed only one (Burns, 1996) made a brief mention. Rodríguez and Niaz (2004b) have reported that of the 43 general physics textbooks analyzed, only one (Ohanian, 1987) described satisfactorily that Millikan's work formed part of a sequence of heuristic principles. Another textbook (Eisberg, 1973) made a simple mention.

In an effort to explore further the presentation of the Millikan oil drop experiment.

Rodríguez and Niaz (2004b) also analyzed 11 physics freshman laboratory manuals. Surprisingly, the manuals do not deal satisfactorily even with the experimental variables that made the experiment so difficult. Just as the textbooks, laboratory manuals attribute Millikan's success to his precise measurements and thus ignore the theoretical rationale behind his experimental data.

It could be argued that the oil drop experiment is too complex and cannot be presented in introductory freshman textbooks. So we decided to study physical chemistry textbooks, which are used in an advanced course that includes atomic structure as a topic. Niaz and Rodríguez (2005) reported that none of the 28 physical chemistry textbooks recognized the controversial nature of the oil drop experiment and ignored the heuristic principles. It is understandable that physical chemistry textbooks may not present the Millikan-Ehrenhaft controversy (it pertains to HPS?) but how do we justify that the experiment could be presented without a satisfactory explanation of the incidence of the experimental variables that made the experiment difficult. This is all the more paradoxical, as physical science is considered by its practitioners to be primarily an experimental science.

Chapter 5

LAWS OF DEFINITE AND MULTIPLE PROPORTIONS IN CHEMISTRY

The French chemist Joseph Louis Proust (1754-1826) was the first to express and provide experimental verification of the law of definite proportions in 1799. The law is generally enunciated in textbooks as: "Different samples of a substance contain its elementary constituents (elements) in the same proportion" (Pauling, 1964, p. 25). The law of multiple proportions is generally attributed to the English chemist John Dalton (1766-1844), who first discovered it in August 1803, according to textbook accounts while working on the composition of the hydrocarbons methane and ethene. Most textbooks enunciate the law as: "When an element combines with another to form more than one compound the masses of the second element combining with a fixed mass of the first element bear a simple ratio to one another" (Taylor, 1942, p. 2).

Some scholars (Thomson, 1825) in the early 19th century popularized the positivist version that Dalton was led to his atomic theory by the discovery of the law of multiple proportions. According to Rocke (1984): "This inductivist version was quite concordant with the then prevalent Victorian model of heroic science" (p. 27). Linus Pauling (1964) clarifies the clarifies the issue by stating categorically: "The discovery of the law of simple multiple proportions was the first great success of Dalton's atomic theory. This law was not induced from experimental results, but was derived from the [atomic] theory and then tested by experiments" (p. 26). Most general chemistry textbooks present the laws of definite and multiple proportions in a simple and uncontroversial manner. In spite of such presentations, a critical appraisal of the history and philosophy of science (HPS) literature shows that the development of the two laws is closely related to the controversial origin of Dalton's atomic theory, starting from the early 19th

century. For example, the laws of definite and multiple proportions were interpreted by some scientists as a manifestation of Gay-Lusaac's empirical law of combining volumes. An alternative interpretation considered the laws to be a manifestation of the atomic nature of matter. These opposing interpretations continued almost to the first decade of the 20th century (Pauling, 1952, pp. 135-136). Textbooks can improve their presentations by including the following heuristic principles (criteria):

Dalton's Atomic Theory Predicted and Explained the Law of Multiple Proportions

The objective of this criterion is to to evaluate if textbooks follow one of the following interpretations with respect to the law of multiple proportions: a) *Inductivist*: Dalton was led to his atomic theory by the discovery of the law of multiple proportions. According to Lakatos (1971) for an inductivist, "... only those propositions can be accepted into the body of science which either describe hard facts or are infallible inductive generalizations from them" (p. 92). In the present case the 'gravimetric combining proportions' would constitute the 'hard facts.' b) *Lakatosian*: This law was not induced from experimental results, but was derived from Dalton's atomic theory and then tested by experiments. Niaz (2001b) has reported that of the 27 general chemistry textbooks analyzed, 4 supported the inductivist interpretation and 16 made no mention to the underlying issues. These latter textbooks, however, followed the approach that scientific theories are preceded by scientific laws, which are in turn preceded by experimental observations. Seven textbooks followed the Lakatosian interpretation and following is an example:

> If a scientific theory is any good, it should lead to predictions about behavior of nature that have not yet been recognized. Reasoning from his theory, Dalton was able to predict a regularity in the weight relations for the case of the same two elements forming two different compounds ... This relationship was borne out by repeated experiments and has come to be called the *law of multiple proportions* (Sienko and Plane, 1971, p. 12).

Dalton's Atomic Theory Predicted and Explained Gay Lusaac's Law of Combining Volumes

The objective of this criterion is to evaluate if textbooks follow one of the following interpretations: a) *Inductivist*: Empirical evidence from Gay-Lusaac's law of combining volumes provided a rationale for accepting the laws of definite and multiple proportions, without the 'superfluous' atomic theory of Dalton. b) *Lakatosian*: Dalton's atomic theory predicted and explained Gay-Lusaac's law of combining volumes. Niaz (2001b) has reported that two textbooks followed the inductivist interpretation. Such presentations emphasized that although Dalton had studied the mass relationships in many reactions, he could find no explanation for Gay-Lusaac's law of combining volumes. None of the textbooks presented the issues involved from a Lakatosian perspective and 25 made no mention of the issues involved. Textbooks generally confused the issues by pointing out that Dalton had not understood Gay-Lusaac's and Avogadro's laws, as he did not accept the existence of diatomic molecules.

Finally, it is not the objective of this study (and the chapter in general) to down play empirical findings, but rather establish a framework in order to understand the relationship between empirical findings, laws and theories. According to Lakatos (1970): "Even then experience still remains, in an important sense, the 'impartial arbiter' of scientific controversy. We cannot get rid of the problem of the 'empirical basis' if we want to learn from experience ..." (p. 131). What we need is not a simple description of the experimental findings, but rather an epistemology based on history and philosophy of science (Matthews, 1994).

Chapter 6

KINETIC THEORY OF GASES

Kinetic theory of gases has been the subject of considerable debate and controversy in the history and philosophy of science literature (Achinstein, 1987, 1991; Brush, 1976; Clark, 1976; De Regt, 1996; Elkana, 1974; Kuhn, 1970; Lakatos, 1970). The role of kinetic theory in understanding gases, liquids, solids, thermodynamics and various other topics has been recognized in the science curricula and textbooks.

CLAUSIUS' SIMPLIFYING (BASIC) ASSUMPTIONS

Clausius' (1857) is considered to be the first full fledged kinetic theory of gases and his following basic assumptions can be considered as a prelude to those of Maxwell's:

(1) Space actually filled by the molecules of the gas must be infinitesimal in comparison to the whole space occupied by the gas;
(2) Duration of the impact (i.e., change of direction) of the molecules must be infinitesimal compared with the time interval between the collisions; and
(3) Influence of the molecular forces between the molecules must be infinitesimal.

Maxwell's Contribution to the Development of the Kinetic Theory of Gases

The starting point of Maxwell's work on the kinetic theory of gases was his reading of the paper by Clausius (1857), entitled: "On the nature of the motion which we call heat" (Garber et al., 1986, p. xix). Similarly Maxwell (1860) recognizes the work of early kinetic theorists, such as Bernouilli, Herapath, Joule and Krönig. Maxwell sets down the following simplifying (basic) assumptions of his theory, which have been summarized by Achinstein in the following terms (1987, p. 410):

1. Gases are composed of minute particles in rapid motion.
2. Particles are perfectly elastic spheres.
3. Particles act on each other only during impact.
4. Motion of the particles is subject to mechanical principles of Newtonian mechanics.
5. Velocity of the particles increases with the temperature of the gas.
6. Particles move with uniform velocity in straight lines striking against the sides of the container, producing pressure
7. Derivation of the distribution law assumes that the x-, y-, and z components of velocity are independent.

According to Achinstein (1987): "How did Maxwell arrive at them [assumptions]? They are highly speculative, involving as they do the postulation of unobserved particles exhibiting unobserved motion" (p. 410). Did Maxwell have an independent warrant (i.e., plausibility of the hypotheses) for his simplifying assumptions? It is plausible to suggest that Maxwell's assumptions are precisely the *ceteris paribus* clauses, which helped him to progress from simple to complex models of the gases. This methodology of Galilean idealization, that is, building of simple to complex models, is an important characteristic of modern non-Aristotelian science (Kitchener, 1993; McMullin, 1985; Niaz, 1993). Lakatos (1970) has endorsed this position: "Moreover, *one can easily argue that ceteris paribus clauses are not exceptions, but the rule in science*" (p. 102). Maxwell's research program is yet another example of a program progressing on inconsistent foundations. Among other assumptions, Maxwell's (1860) paper was based on 'strict mechanical principles' derived from Newtonian mechanics and yet at least two of Maxwell's simplifying assumptions (referring to movement of particles and consequent generation of pressure) were in contradiction with Newton's hypothesis explaining the gas laws based on repulsive forces between particles. Newton provided one of the first explanations of Boyle's law in his *Principia* (1687) in the following terms: "If a gas is composed of particles that exert repulsive forces on their neighbors, the magnitude of force being inversely as the distance, then the pressure will be inversely as the volume" (Brush, 1976, p. 13). Due to Newton's vast authority, Maxwell even in his 1875 paper reiterated that Newtonian principles were applicable to unobservable parts of bodies (Achinstein, 1987, p. 418). Brush (1976) has pointed out the contradiction explicitly: ". . . Newton's laws of mechanics were ultimately the basis of the kinetic theory of gases, though this theory had to compete with the repulsive theory attributed to Newton" (p. 14).

A Lakatosian Interpretation of Maxwell's Research Program

According to Lakatos (1970), the *negative heuristic* represents the 'hard core' of the research program, consisting of simplifying (basic) assumptions considered 'irrefutable' by the decision of the community of scientists. The *positive heuristic* represents the construction of a 'protective belt' consisting of a ". . . partially articulated set of suggestions or hints on how to change, develop the 'refutable variants' of the programme" (Lakatos, 1970, p. 135). The scientist lists anomalies, but as long as his research program sustains its momentum, he may freely put them aside, and "*It is primarily the positive heuristic of his programme, not the anomalies, which dictate the choice of his problems*" (Lakatos, 1970, p. 99). A research program is progressing if it frequently succeeds in converting anomalies into successes, that is, explainable by the theory – referred to as '*progressive problemshifts*'.

Based on Lakatos' (1970) philosophy of science, Clark (1976) considers the 7 simplifying assumptions as the hard core (negative heuristic) of Maxwell's research program and summarizes it in the following terms: ". . . the behaviour and nature of substances is the *aggregate* of an enormously large number of very small and constantly moving elementary individuals subject to the laws of mechanics" (p. 45). Similarly, Clark (1976, p. 45) considers the following methodological directives as the positive heuristic of Maxwell's research program: (1) Make specific assumptions as to the nature of the elementary particles and as to their available degrees of freedom subject to the laws of mechanics; (2) All interactions shall be treated according to the laws of mechanics, while distribution of the properties of the molecular motion among the molecules shall be treated according to the laws of statistics; (3) Try to weaken or if possible eliminate the simplifying assumptions, so as to simulate conditions obtaining in a 'real' gas.

'Progressive Problemshifts' in Maxwell's Research Program

Maxwell's major contribution was to make predictions beyond the hydrodynamical laws (Boyle, Charles, Gay-Lusaac, etc.), referring to transport properties of gases. In subsequent work, Maxwell (1965) and others reduced/modified their original simplifying assumptions as formulated in the positive/negative heuristic of the program, in order to obtain:

(a) A more rigorous deduction of the law of velocities in a steady state;
(b) A better approximation of the effect of molecular collisions upon the values of transport coefficients (Clark, 1976, p. 54);
(c) Instead of considering the particles as 'elastic spheres' he introduced the concept of 'centers of force', considered to be a 'progressive problemshift';
(d) In 1875, although Maxwell still maintained that Newtonian principles were applicable to understand the behavior of gases, he recognized the contradictory nature of one of his basic assumptions by pointing out that, "The pressure of a gas cannot therefore be explained by assuming repulsive forces between the particles" (Maxwell, 1875, p. 422);
(e) Boltzmann was particularly responsible for eliminating simplifying assumptions in Maxwell's theory.

Chapter 7

VAN DER WAALS' EQUATION OF STATE: A 'PROGRESSIVE PROBLEMSHIFT'

Development of van der Waals' (1873) thesis (continuity of the gaseous and liquid state) was based on the kinetic theory of gases, Clausius' virial theorem and the experiments of Joule and Thomson, which showed that the temperature of a gas lowers as it expands. Van der Waals reasoned that the inter-molecular forces which account for the cohesion of the liquid phase represented a property of the molecular model, and hence their effect should still be appreciable in the gaseous phase. It is interesting to observe that if Maxwell's simplifying assumptions were speculative (Achinstein, 1987), van der Waals in a sense followed the same methodology, viz., "... without elaborate justification. Its appropriateness should be judged by van der Waals' results rather than by a priori arguments" (Gavroglu, 1990, p. 219). Using his equation of state, van der Waals, precisely reproduced Andrews' (1869) experimental results (isotherms for carbon dioxide), which demonstrated the continuity of the transition from the gaseous to the liquid state. The importance of van der Waals work can be understood better if we consider his contribution as an attempt to reduce/modify the simplifying assumptions of Maxwell's theory. Clark (1976) interprets van der Waals contribution in the following terms: "*What is important is that the novel predictions were found to be in good agreement with experiment. Application of the heuristic of the kinetic programme had resulted in empirical growth*, ['progressive problemshift'] in this case the discovery of a *new general law*" (p. 60).

Chapter 8

KINETIC THEORY AND CHEMICAL THERMODYNAMICS AS RIVAL RESEARCH PROGRAMS

It is important to note that although in retrospect, these days we recognize the kinetic theory as one of the greatest achievements of 19th century science it was subject to considerable criticism by leading scientists of the day. Ostwald (1927) criticized the kinetic theory for it's ". . . superficial habit to cover up rather than promote actual scientific tasks by *arbitrary assumptions* about atomic positions, motions and vibrations" (p. 178, emphasis added). It is interesting to observe, that the very *simplifying assumptions* that helped Maxwell and recognized as the 'hard core' of the research program by modern philosophers of science, were considered to be arbitrary by some influential 19th century scientists (especially E. Mach and W. Ostwald). It is not difficult to appreciate that the 'hard core' of the kinetic theory's research program had a certain background to it, and to consider it as 'arbitrary' was indeed a superficial criticism. Duhem (1962), another important critic, questioned the atomic models used by the kinetic theory and critically appraised it with respect to chemical thermodynamics: "Thermodynamics had reached maturity and constitutional vigour when the kinetic hypothesis came along to bring it assistance it did not ask for, with which it has nothing to do and to which it owed nothing" (p. 95). The opposition of many of these critics has been attributed to their philosophical approach to science: "... the rise of the school of 'Energetics' [thermodynamics] championed by Mach and Ostwald, represents an early attempt of the positivist philosophy to limit the scope of science. This school held that to use modern terminology the atom was not an 'observable', and that physical theories should not, therefore, make use of the concept" (Jaynes, 1967, p. 80). Brush (1976) has reasoned in a similar vein: "Those scientists who did suggest that the kinetic theory be abandoned in the later 19th century did so not because of empirical difficulties but because of a more deep seated purely philosophical objection. For those who believed in a positivist methodology, any theory based on invisible and undetectable atoms was unacceptable" (p. 1169). Brush (1976) goes on to emphasize: "The leaders of this reaction, in the physical sciences, were Ernest Mach, Wilhelm Ostwald, Pierre Duhem, and Georg Helm. Mach recognized that atomic hypotheses could be useful in science but insisted, even as late as 1912, that atoms must not be considered to have a real existence ... they denied that kinetic theories had any value at all, even as hypotheses" (p. 245). Einstein's criticism of Mach and Ostwald's philosophical views clarifies the issues even further: "The prejudices of these scientists against the atomic theory can be undoubtedly attributed to their positivistic philosophical views. This is an interesting example of how philosophical prejudices hinder a correct interpretation of facts even by scientists with bold thinking and subtle intuition" (Quoted by Suvorov, 1966, p. 578).

Chapter 9

UNDERSTANDING THE BEHAVIOR OF GASES: FROM HYDRODYNAMIC LAWS TO KINETIC THEORY

The major contribution of the kinetic theory scientists (Clausius, Maxwell and Boltzmann) was to have gone beyond the observational, hydrodynamic laws (Boyle, Charles, Gay-Lusaac) and predict the internal properties of gases. From an epistemological perspective the original postulation of the Ideal Gas Law (1660 onwards) based on the experimental data (i.e., manipulating particular variables: P, V, n, and T) of Boyle, Charles, Gay-Lusaac and others can be considered as primarily an inductive process. On the other hand, the later development of the law based on the kinetic theory of Maxwell and Boltzmann (1860 onwards) comes quite close to some form of scientific idealization (Lakatos, 1970). It seems plausible to suggest that resolution of gas problems based on the Ideal Gas Law, derived by the inductive process, primarily requires manipulation of the different variables of the equation ($PV = nRT$) and thus can be characterized by the 'algorithmic mode' (Niaz and Robinson, 1992). On the other hand, resolution of gas problems based on the Ideal Gas Law, which derives its meaning from the kinetic theory of Maxwell requires the understanding of a model (pattern) within which data appear intelligible, that is, a sort of 'conceptual gestalt' (Hanson, 1958, p. 90

Chapter 10

FROM 'ALGORITHMIC MODE' TO 'CONCEPTUAL GESTALT' IN STUDENTS' UNDERSTANDING OF THE BEHAVIOR OF GASES

Based on the previous chapters it is plausible to suggest that students' understanding of the behavior of gases goes through solving algorithmic problems (based on experimental hydrodynamical properties), and later conceptual understanding based on the kinetic theory. A review of the literature shows how freshman students generally perform fairly well on algorithmic problems and have considerable difficulty in solving conceptual problems based on an understanding of the kinetic theory (Niaz and Robinson, 1992).

Chapter 11

CRITERIA FOR EVALUATION OF TEXTBOOKS

Based on the previous chapters it is plausible to suggest that in order to understand kinetic theory, textbooks need to refer to the following heuristic principles (criteria):

MAXWELL'S SIMPLIFYING (BASIC) ASSUMPTIONS

Niaz (2000b) has shown that of the 22 general chemistry textbooks analyzed, 17 simply mentioned that the postulates of the kinetic theory were 'assumptions' and only 3 described Maxwell's simplifying assumptions satisfactorily and following is an example:

> At this point we want to build a model (theory) to explain *why* a gas behaves as it does . . . laws do not tell us *why* nature behaves the way it does. Scientists try to answer this question by constructing theories (building models). The models in chemistry are speculations about how individual atoms or molecules ... cause the behavior of macroscopic systems (collections of atoms and molecules in large enough numbers so that we can observe them). A model is considered successful if it explains known behavior and predicts correctly the results of future experiments. But a model can never be proved absolutely true. In fact, by its very nature *any model is an approximation* and is doomed to fail at least in part. Models range from the simple (to predict approximate behavior) to the extraordinarily complex (to account precisely for observed behavior) . . . A relatively simple model that attempts to explain the behavior of an ideal gas is the *kinetic molecular theory*. This model is based on speculations about the behavior

of the individual particles (atoms or molecules) in a gas (Zumdahl, 1993, pp. 434–435).

This presentation emphasizes speculative models and that models develop (tentativeness) in order to explain the behavior of gases. It can be argued that such presentations are not based on an overt understanding of history and philosophy of science. Nevertheless, most teachers would agree, that it constitutes a logical and helpful preamble before presenting the postulates of the kinetic theory. Rodríguez and Niaz (2004c) have reported that of the 30 general physics textbooks analyzed only two presented satisfactorily the role played by simplifying assumptions and five made a simple mention.

INCONSISTENT NATURE OF MAXWELL'S RESEARCH PROGRAM

All the general chemistry textbooks ignored that Maxwell's program, although successful, was also based on an inconsistent foundation. Many textbooks explicitly invoke Newtonian mechanics along with Maxwell's presentation of the kinetic theory, without realizing an inherent contradiction. Similarly, Rodríguez and Niaz (2004c) have reported that none of the general physics textbooks made a satisfactory presentation.

VAN DER WAALS' CONTRIBUTION --- REDUCING / MODIFYING BASIC ASSUMPTIONS

Niaz (2000b) has reported that 14 general chemistry textbooks described satisfactorily van der Waals' contribution as an attempt to reduce/modify the basic assumptions. Following is an example of a satisfactory description:

> Van der Waals examined critically the postulates of the kinetic molecular theory and recognized that some of these postulates had to be modified if the kinetic theory was to account more accurately for the behavior of real gases (Quagliano and Vallarino, 1969, p. 140).

Most textbooks do not conceptualize van der Waals' contribution as an attempt to modify Maxwell's simplifying assumptions, which led to a 'progressive problemshift.' Brush (1976) has highlighted this aspect: "Thus, 100 years after its original publication, van der Waals' theory still serves as an illuminating example of how an astute scientist can penetrate to the heart of an important but complex phenomenon by the proper choice of *simplifying approximations*, thereby opening up a new field of theoretical and experimental research" (p. 251, emphasis added). Similarly, Rodríguez and Niaz (2004c) have reported that only one general physics textbook made a satisfactory presentation and 5 simply mentioned it.

KINETIC THEORY AND CHEMICAL THERMODYNAMICS AS RIVAL RESEARCH PROGRAMS

Niaz (2000b) has reported that none of the general chemistry textbooks described this rivalry satisfactorily and only three made a simple mention. Textbooks present the development of kinetic theory as a smooth process that involved no intellectual conflicts.

UNDERSTANDING BEHAVIOR OF GASES --- FROM ALGORITHMIC TO CONCEPTUAL

None of the textbooks described satisfactorily or briefly mentioned the two modes of solving gas problems, viz., the algorithmic mode and that of conceptual understanding (Niaz, 2000b). Textbooks do not visualize the transition from hydrodynamical laws (e.g., Boyle) to kinetic theory (Maxwell, Boltzmann), as an opportunity to increase our conceptual understanding of gaseous behavior. Some textbooks did include problems that could be considered as conceptual, however they lacked the framework based on the two modes of understanding gases. Similarly, Rodríguez and Niaz (2004c) found that none of the general physics textbooks explicitly recognized the difference between the two modes of solving gas problems.

Finally, these study show that most textbooks ignore an essential aspect of scientific progress, viz., how the diversity of elements found in the behavior of gases are studied/controlled through *ceteris paribus* clauses (approximations), leading to tentative theories that increase in their explanatory/heuristic power. Even when textbooks present historical details it invariably is in the form of names of famous scientists including their pictures, year of work, and anecdotes.

It appears that textbooks ignore historical details, not due to limitations of space but rather due to a lack of a history and philosophy of science framework.

Chapter 12

COVALENT BOND

Covalent bonding is considered to be a difficult topic for freshman students. Most textbooks and chemistry teachers would agree that the simplest approach to bonding is in terms of Lewis diagrams and the concept of the shared pair electron. This consensus leads precisely to a dilemma. In an attempt to simplify the topic most textbooks present rules (algorithms) for writing Lewis diagrams for covalent bonds, which are memorized by the students. Such presentations do not lead the students towards conceptual understanding of the difference between covalent (sharing of electrons) and ionic (transfer of electrons) bonds. Recent research in science education has shown an increasing interest in emphasizing conceptual understanding (Niaz and Robinson, 1992, 1993).

A HISTORY AND PHILOSOPHY OF SCIENCE PERSPECTIVE

Lewis (1916) is generally considered to have presented the first satisfactory model of the covalent (shared paired) bond based on the cubic atom. The genesis of the cubic atom can be traced to an unpublished memorandum written by Lewis in 1902 and recounted by him in the following terms:

> In the year 1902 (while I was attempting to explain to an elementary class in chemistry some of the ideas involved in the periodic law) becoming interested in the new theory of the electron (Thomson's discovery of the electron in 1897), and combining this idea with those which are implied in the periodic classification, I formed an idea of the inner structure of the atom (model of the cubic atom) which, although it contained crudities, I have ever since regarded as

representing essentially the arrangement of the electrons in the atom (Lewis, 1923, pp. 29-30).

Lewis (1916) reproduced the six postulates of his 1902 theory of the cubical atom at length of which the third postulate was the most surprising: "The atom tends to hold an even number of electrons in the shell, and especially to hold eight electrons which are normally arranged symmetrically at the eight corners of a cube"(p. 768). This was the most controversial feature of Lewis's theory and what was the warrant for including this postulate, which later led to the formulation of the 'rule of eight' or the 'octet rule.' Lewis postulated that the eight electrons of an octet formed the eight corners of a cube, as this provided, "... the most stable condition for the atomic shell" (p. 774). Thus the single bond was conceived of as two cubic atoms with a shared edge (pair of electrons) and the double bond as two cubes with a common face.

Lewis's Model of Sharing Electrons (Covalent Bond) as a Rival to the Transfer of Electrons (Ionic Bond)

From a philosophy of science perspective the rivalry between competing theories (paradigms / research programs) is an integral part of scientific progress. Kohler (1971) has presented a detailed account of the origin of Lewis's ideas:

> When it was first proposed, Lewis's theory was completely out of tune with established belief. For nearly 20 years it had been almost universally believed that all bonds were formed by the complete transfer of *one* electron from one atom to the other. The paradigm was the ionic bond of Na^+ Cl^-, and even the bonds in compounds such as methane or hydrogen were believed to be polar, despite their lack of polar properties. From the standpoint of the polar theory *the idea that two negative electrons could attract each other or that two atoms could share electrons was absurd*" (p. 344).

It is important to note that a major proponent of the rival ionic bond was none other than the leading British scientist J.J. Thomson (1907). He accepted that overlapping of corpuscles could produce a non-polar bond in theory, still in reality all bonds were polar.

Origin of the Covalent Bond: A Baconian Inductive Ascent?

A major premise of historians who follow the Baconian inductive ascent is that scientific theories and laws are primarily driven by experimental observations. Such empiricist interpretations consider scientific progress to be dichotomous, viz., experimental observations lead to scientific laws, which later facilitate the elaboration of explanatory theories. Conceptualization of the covalent bond by chemists and textbooks approximates quite closely to a Baconian inductive ascent, according to the following stages:

1. The finding that diatomic molecules such as H_2 and the hundreds of compounds found by organic chemists in the late nineteenth century could not be understood by the ionic bond.
2. Postulation of the non-polar shared pair covalent bond by Lewis as an inductive generalization.
3. Theoretical explanation offered by quantum theory (Pauli's exclusion principle) as to how two electrons (in spite of the repulsion) can occupy the same space.

The third stage has been corroborated by the following interpretation of the events by Rodebush (1928):

> It will be recognized by the chemist however that Pauli's rule [exclusion principle] is only a short hand way of saying what Lewis has assumed for many years as the basis of his magnetochemical theory ... of valence. If the electrons are paired in the atom magnetically, it is easy to see how two unpaired electrons in different atoms may be coupled magnetically and form the nonpolar bond (p. 516).

The crux of the issue is that the inductivist interpretation construes Pauli's exclusion principle as the theoretical explanation and ignores the fact that Lewis's cubic atom was crucial for his later explanation of the sharing of electrons. Thus scientific progress is characterized by a series of theories or models (cubic atom to exclusion principle), which vary in the degree to which they explain experimental findings.

CRITERIA FOR EVALUATION OF GENERAL CHEMISTRY TEXTBOOKS

Based on the history and philosophy of science perspective it is plausible to suggest that textbooks need to refer to the following heuristic principles (criteria).

Lewis's Cubic Atom as a Theoretical Device for Understanding the Sharing of Electrons

Niaz (2001c) has reported that of the 27 general chemistry textbooks analyzed only three described satisfactorily the role played by Lewis's cubic atom and following is an example:

> Lewis assumed that the number of electrons in the outermost cube on an atom was equal to the number of electrons lost when the atom formed positive ions ... he assumed that each neutral atom had one more electron in the outermost cube than the atom immediately preceding it in the periodic table. Finally, he assumed it took eight electrons --- an octet --- to complete a cube. Once an atom had an octet of electrons in its outermost cube, this cube became part of the core, or kernel, of electrons about which the next cube was built (Bodner and Pardue, 1989, p. 273).

With this introduction, authors explain the formation of the covalent bond:

> By 1916, Lewis had realized that there was another way atoms could combine to achieve an octet of valence electrons --- they could share electrons. Two fluorine atoms, for example, could share a pair of electrons and thereby form a stable F_2 molecule in which each atom had an octet of valence electrons (p. 274).

Authors provide pictures of the individual cubes of fluorine coming together to share an edge and thus form the covalent bond in which the eight electrons are oriented towards the corners of a cube. Furthermore, the authors reproduce Lewis's 1902 memo with hand written drawings of the cubic atom that was included by Lewis (1923) in his now famous book on valence.

Sharing of Electrons (Covalent Bond) had to Compete with the Transfer of Electrons (Ionic Bond)

One of the textbooks mentioned and none described satisfactorily that Lewis's idea of sharing electrons had to compete with the transfer of electrons (Niaz, 2001c). A number of textbooks start with the presentation of the covalent bond in terms that can be useful. Nevertheless, these authors do not interpret the origin of the covalent bond as a rival research program and following is an example:

> The vast majority of chemical substances do not have the characteristics of ionic materials; water, gasoline, banana peelings, hair, antifreeze, and plastic bags ... For the very large class of substances that do not behave like ionic substances, we need a different model for the bonding between atoms. Lewis reasoned that an atom might acquire a noble-gas electron configuration by sharing electrons with other atoms. A chemical bond formed by sharing a pair of electrons is called a *covalent bond* (Brown and LeMay, 1988, p. 233).

A brief reference to the historical details can facilitate conceptual understanding and make students aware of the difficulties involved and how scientists (like students) resist change.

Covalent Bond: Inductive Generalization / Derived from the Cubic Atom

The objective of this criterion was to evaluate if the textbooks follow one of the following interpretations with respect to the origin of the covalent bond: a) Inductivist: Lewis's covalent bond was an inductive generalization based on: stability of the noble gases or formation of the hydrogen molecule leads to a lowering of the energy or Helium an inert gas has a pair of electrons; b) Lakatosian: Lewis's covalent bond was not induced from experimental evidence but derived from the cubic atom. Niaz (2001c) has reported that 23 textbooks considered the origin of the covalent bond to be an inductive generalization and following is an example:

> Because all noble gases (except He) have eight valence electrons, many atoms undergoing reactions also end up with eight electrons. This observation had led to what is known as the *octet rule* (Brown and LeMay, 1988, p. 225).

Such presentations are quite representative of most textbooks and show clearly that the octet rule is sustained by empirical evidence. The alternative interpretation (Lakatosian) would have emphasized the cubic atom --- a hypothetical entity. Only two textbooks (Bodner and Pardue, 1989; Mahan and Myers, 1987) presented the Lakatosian interpretation. These textbooks trace the origin of the stability of the covalent bond to the cubic atom and go to considerable length to show that Lewis's ideas developed slowly based on conjectures and were tentative. At first sight the difference between the two approaches may seem trivial. Nevertheless, the inductivist approach considers all scientific findings to be driven by experiment, which comes quite close to a 'Baconian inductive ascent' (Niaz, 1999). It is important to note that it was Lewis's model of the cubic atom and not the octet rule, which provided the first tentative (conjectural) explanation of the covalent bond. A better explanation ('progressive problemshift') is provided by Pauli's exclusion principle (Pauli, 1925), subject of the next criterion.

Pauli's Exclusion Principle as an Explanation of the Sharing of Electrons in Covalent Bonds

Niaz (2001c) has reported that five textbooks mentioned and eight described satisfactorily Pauli's principle as an explanation of the sharing of electrons (with opposite spin) in covalent bonds and following is an example:

> When two hydrogen atoms form a covalent bond, the atomic orbitals overlap in such a way that the electron clouds reinforce each other in the region between the nuclei, and there is an increased probability of finding an electron in this region. According to Pauli's exclusion principle, the two electrons of the bond must have opposite spins. The strength of the covalent bond comes from the attraction of the positively charged nuclei for the negative cloud of the bond...(Mortimer, 1983, p. 135).

Finally, a historical reconstruction shows that the Lewis's cubic atom was the first attempt to explain the stability of the covalent bond and later Pauli's exclusion principle provided further support. Transition from Lewis's cubic atom → Pauli's exclusion principle → what next, provides an illustration of how scientific knowledge is tentative.

Chapter 13

PERIODIC TABLE

Most chemistry teachers consider the periodic table to be an important concept, both in principle and practice. It facilitates a succinct organization and understanding of the fundamental building blocks of chemistry, the chemical elements (Atkins, 1995). According to one historian of chemistry the periodic table, ". . . has contributed much more than mere classification. It has been a conceptual tool which has predicted new elements, predicted unrecognized relationships, served as a corrective device, and fulfilled a unique role as a memory and organization device" (Ihde, 1969, p. ix). Mendeleev's periodic table formed part of his textbook (Principles of Chemistry, first written between 1868 and 1870), in which he endeavored to facilitate students' understanding. In spite of the long history of the periodic table and its relevance for chemistry and chemistry education, historians and philosophers of science are still trying to understand its origin, nature, and development (Allchin, 2005; Bensaude-Vincent, 1986; Brush, 1996, 2005, 2007; Lipton, 2005).

A HISTORY AND PHILOSOPHY OF SCIENCE FRAMEWORK

Role of Accommodation and Prediction in Development of the Periodic Table

Mendeleev (1869) enunciated the first form of his periodic law and later elaborated in the following terms: "The properties of simple bodies, the constitution of their compounds, as well as the properties of these last, are periodic functions of the atomic weights of elements" (Mendeleev, 1879, p. 267).

Elucidation of the concept of atomic weight by Stanislao Cannizaro at Karlsruhe was crucial in the discovery of the periodic law. Elaboration of the periodic table was difficult and took a long time due to lack of a definite conception of atomic weight, which is very closely connected with the definitions of molecules and atoms. Availability of the atomic weights of about 60 elements enabled Mendeleev to accommodate the elements in the table according to various physicochemical properties (density, specific heat, atomic weight, atomic volume, melting point, valence, oxides, chlorides, and sulfides). Historians and philosophers of science continue to debate as to what was crucial for the acceptance of the periodic law: accommodation of the existing elements or the prediction of new ones. Lipton (1991, 2005) and Maher (1988) favor a predictivist thesis, viz., Mendeleev's law was accorded a greater recognition after the discovery of the first predicted element (gallium) in 1875. Allchin (2005) and Brush (1996, 2005, 2007) seem to suggest that accommodations were more important.

Mendeleev left various vacant spaces in his table and made many predictions and, of these, the following are the most important:

a) Eka-aluminum (atomic weight=68, density=6.0, atomic volume=11.5). This was discovered by the French chemist Paul Emile Lecoq de Boisbaudran in 1875, and was named gallium.
b) Eka-boron (atomic weight=44, density=3.5). This was discovered by the Swedish chemist Lars-Frederik Nilson in 1879, and was named scandium.
c) Eka-silicon (atomic weight=72, density=5.5, atomic volume=13). This was discovered by the German chemist Clemens Alexander Winkler in 1886, and was named germanium.

Besides the atomic weights and physical properties, some of the chemical properties (formation of oxides, chlorides) of the predicted elements coincided to a remarkable degree with the discovered elements. According to van Spronsen (1969), after the discovery of gallium in 1875, ". . . Mendeleev rightly concluded that the validity of the periodic system of elements could no longer be questioned. The confirmation of this prediction may certainly be called the culminating point in the history of the periodic system" (p. 221). This precisely is the point of contention among philosophers of science, viz., what made Mendeleev's periodic law valid—accommodations dating from 1869 or the predictions from 1875 onwards. According to Brush (1996), scientists generally propose a hypothesis, deduce its consequences, make predictions, and do experiments to see if the predictions are borne out. Ziman (1978) believes that the ". . . fundamental purpose of science is to acquire the means for reliable prediction" (p. 32). Actual scientific practice, however, is much more complex and controversial.

Periodicity in the Periodic Table as a Function of the Atomic Theory

According to van Spronsen (1969):

> The actual development of the periodic system seemed to require a catalyst! We think it proper to attribute this catalytic action to Cannizaro's famous Karlsruhe lecture at the 1860 Congress. He *made the distinction between atoms and molecules and defined such concepts as valence*; this initiated the second stage of the discovery and started the history proper of the periodic system of chemical elements (p. 1).

In spite of this fairly categorical statement with respect to the role played by the atomic theory by a major historian of the periodic table, it is still possible to observe that some historians and textbooks attribute the success primarily to empirically observed properties of the elements (inductive generalization). Many chemistry students must have wondered as to how Mendeleev and the other co-discoverers could have conceptualized the underlying theoretical rationale of the elements that manifested itself in periodicity. It is important to recall that most of the pioneering work of Mendeleev was conducted from 1869 to 1889, before Thomson (1897), Rutherford (1911), Bohr (1913), and Moseley (1913) laid the foundations of the modern atomic theory. So how could Mendeleev conceptualize periodicity as a function of the atomic theory? An answer to this question will precisely show Mendeleev's ingenuity, far-sightedness, creativity, and the ability to 'speculate.' Despite Mendeleev's own ambivalence and ambiguity, a historical reconstruction does provide a convincing story of this remarkable contribution to our knowledge.

Before presenting the reconstruction it is important to note that Mendeleev had the following important sources of information: Dalton's atomic theory; law of multiple proportions; Cannizaro's Karlsruhe lecture; fairly reliable atomic weights; atomicity (valence); and various physical and chemical properties of the elements.

- *Stage 1.* In his first publication, Mendeleev (1869) referred to the relationship between periodicity, atomic weights, and valence: "The arrangement according to atomic weight corresponds to the *valence* of the element and to a certain extent the difference in chemical behavior, for example Li, Be, B, C, N, O, F." (p. 405).
- *Stage 2.* After the discovery of gallium and scandium, Mendeleev expressed the relationship between atomic weight and atomic theory much more explicitly:

> It is by studying them [atomic and molecular weights], more than by any other means, that we can *conceive the idea of an atom and of a molecule*. By this fact alone we are enabled to perceive the great influence that studies carried on in this direction can exercise on the progress of chemistry ... The expression atomic weight* implies, it is true, the hypothesis of the atomic structure of bodies. (Mendeleev, 1879, p. 243, emphasis added)

The asterisk leads the reader to the following footnote: "By replacing the expression of atomic weight by that of elementary weight, I think we should, in the case of elements, avoid the conception of atoms." The footnote shows Mendeleev's ambiguity/ambivalence toward the atomic theory and will be dealt with later (see Stage 6).

- *Stage 3.* Another example of Mendeleev's ambivalence can be observed from the following: "I shall not form any hypotheses, either here or further on, to explain the nature of the periodic law; for, first of all, the law itself is too simple* "(Mendeleev, 1879, p. 292). The asterisk leads the reader to the following footnote: "However, I do not ignore that to completely understand a subject we should possess, independently of observations [and experiences] and of laws [as well as systems], the meanings of both one and the other."
- *Stage 4.* Although Mendeleev stated in 1879 that he would not formulate a hypothesis,10 years later in his famous Faraday lecture, Mendeleev (1889) not only attributed the success of the periodic law to Cannizaro's ideas on the atomic theory (pp. 636–637) but went on to explicitly formulate the following hypothesis: " ... the veil which conceals the true conception of mass, it nevertheless indicated that the explanation of that conception must be searched

for in the masses of atoms; the more so, *as all masses are nothing but aggregations, or additions, of chemical atoms ...*" (p. 640).
- *Stage 5.* Again, at the Faraday lecture, Mendeleev (1889) took extreme care to explain the periodicity of properties of chemical elements on the basis of atomic theory:

The periodic law has shown that our chemical individuals [atoms] display a harmonic periodicity of properties, dependent on their masses .... An example will better illustrate this view. The atomic weights—

Ag =108 Cd = 112 In = 113 Sn = 118 Sb = 120 Te = 125 I = 127

steadily increase, and their increase is accompanied by a modification of many properties which constitutes the essence of the periodic law. Thus, for example, the densities of the above elements decrease steadily, being respectively—

10.5 8.6 7.4 7.2 6.7 6.4 4.9

while their oxides contain an increasing quantity of oxygen—

Ag_2O Cd_2O_2 In_2O_3 Sn_2O_4 Sb_2O_5 Te_2O_6 I_2O_7

But to connect by a curve the summits of the ordinates expressing any of these properties would involve the rejection of Dalton's law of multiple proportions. Not only are there no intermediate elements between silver, which gives AgCl, and cadmium which gives $CdCl_2$, but, according to the very essence of the periodic law there can be none; in fact a uniform curve would be inapplicable in such a case, as it would lead us to expect elements possessed of special properties at any point of the curve (pp. 640–641).

This is a clear acknowledgment of the role played by the atomic theory to explain periodicity in the periodic table. Mendeleev clearly conceptualized the relationship between the discontinuous function of the periodic properties and its dependence on the law of multiple proportions, which in the ultimate analysis meant atomic theory. To support our claim we once again quote from Mendeleev's Faraday lecture:

the periodic law has clearly shown that the masses of the atoms increase abruptly, by steps, which are clearly connected in some way with *Dalton's law of multiple proportions*; ... While connecting by new bonds the theory of the chemical elements with Dalton's theory of multiple proportions, or atomic structure of bodies, the periodic law opened for natural philosophy a new and wide field for speculation. (Mendeleev, 1889, p. 642)

Interestingly, Mendeleev even seems to be considering the law of multiple proportions synonymous with Dalton's atomic theory.

- *Stage 6.* At this stage we to refer to Mendeleev's ambiguity/ambivalence toward the atomic theory. Throughout the 19th century, positivism was the dominant philosophy, which led all scientific work to be based strictly on experimental observations and all hypothetical propositions were considered speculative and hence nonscientific (Brush, 1976; Gavroglu, 2000; Holton, 1992). Mendeleev was clearly aware of this and on many occasions went out of his way to emphasize that the periodic ". . . law itself was a legitimate induction from the verified facts"(Mendeleev, 1889, p. 639). Mendeleev emphasized the inductive aspect of the periodic law in the light of the anti-atomist Marcellin Berthelot's (1827–1907) criticism, ". . . the illustrious Berthelot, in his work Les origins de l' Alchimie, 1885, 313, has simply mixed up the fundamental idea of the law of periodicity with the ideas of Prout, the alchemists, and Democritus about primary matter. But the periodic law, based as it is on the solid and wholesome ground of experimental research, has been evolved independently of any conception as to the nature of the elements; …" (Mendeleev, 1889, p. 644). Mendeleev's dilemma was that, on the one hand, he could rightly claim that the periodic law was based on experimental properties of the elements (an aspiration of scientists in the late 19th century), and yet he could not give up the bigger challenge, viz., the possible causes of periodicity, and hence importance of atomic theory.

Should Mendeleev's Contribution be Considered a Theory or an Empirical Law?

There seems to be considerable controversy among philosophers of science with respect to the nature of Mendeleev's contribution. Wartofsky (1968) clearly considers Mendeleev's contribution to be more than a simple empirical law:

Mendeleev, for example, predicted that the blank space of atomic number 32, which lies between silicon and tin in the vertical column, would contain an element which was grayish-white, would be unaffected by acids and alkalis, and would give a white oxide when burned in air, and when he predicted also its atomic weight, atomic volume, density and boiling point, he was using the periodic table as a hypothesis from which predictions could be deduced (p. 203).

Ziman (1978) recognizes the importance of predictions with respect to the validity of a theory, and hence Mendeleev's contribution can be considered as a theory:

> Needless to say, the most impressive way of validating a scientific theory is to confirm its *predictions* the persuasive power of a successful prediction arises from the fact that it could not have been deliberately contrived. The most famous examples, such as Mendeléef's prediction of the existence and properties of undiscovered elements to fill the gaps in the periodic table, have astonishing rhetorical power (p. 31).

Shapere (1977) refers to the fact that, historically, Mendeleev's work has been referred to as a classification, system, table, or a law. Nevertheless, in his opinion, the periodic table is neither a law nor a theory but rather an ordered domain.

Bensaude-Vincent (1986) suggests that Mendeleev:

> was able to accomplish the positivist ideal for a mature science: to summarize all the known facts and laws in a systematic table; . . . Mendeleev belonged to a strict positivist tradition: his rejection of all hypotheses on the origin of the elements, his search of a single general law gathering the largest number of chemical data, his practice of classification, are all typical attitudes of the 'esprit positif' according to A. Comte (p. 14).

This attribution of 'esprit positif' to the work of Mendeleev, however, contrasts sharply with what Mendeleev had to say about his contribution:

> If statements of fact themselves depend upon the person who observes them, how much more distinct is the reflection of the personality of him who gives an account of methods and of philosophical speculations which form the essence of science! For this reason there will inevitably be much that is subjective in every objective exposition of science (Preface to the sixth Russian edition, reproduced in Mendeleev, 1897, p. vii).

It is of interest to observe how the conceptualizations of Wartofsky (1968) and Ziman (1978) coincide on the one hand, and the degree to which they differ with those of Bensaude-Vincent (1986) and Shapere (1977). In contrast, Lakatos (1970), and perhaps (!) Mendeleev (1897) conceptualize the problem in an entirely different framework, which in our opinion is quite helpful in understanding Mendeleev's periodic table. We do not necessarily have to follow the law/theory dichotomy, but rather it is plausible to suggest that Mendeleev's work can be considered as an 'interpretative' theory, which became 'explanatory' (cf. Lakatos, 1970) after the periodic law was based on atomic numbers (Bohr, 1913; Moseley, 1913; Thomson, 1897). These considerations, if included in the textbooks, can facilitate students' understanding with respect to how scientific progress is laden with controversies, contradictions, and alternative interpretations.

EVALUATION OF GENERAL CHEMISTRY TEXTBOOKS

Based on the history and philosophy of science framework presented in the previous section, we present here criteria (heuristic principles) for the evaluation of general chemistry textbooks:

Importance of Accommodation in the Development of the Periodic Table

Brito, Rodríguez and Niaz (2005) have reported that of the 57 general chemistry textbooks analyzed, 55 presented a satisfactory description of the importance of accommodation of the chemical elements according to their physicochemical properties in the periodic table. One textbook gave the following advice to the students: "*It is extremely important for you to connect the configuration of an element and its position in the periodic table, since this will allow you to organize a large number of chemical facts*" (Kotz and Purcell, 1991, p. 325, original italics). This shows that the textbooks are fully aware of the role played by accommodation in the development of the periodic table.

Importance of Prediction as Evidence to Support the Periodic Law

Thirty textbooks emphasized the importance of prediction satisfactorily as evidence to support the periodic law and, of these, 29 textbooks compared the properties of at least one of the predicted elements (Ga, Sc, and Ge) with the experimental values. This comparison was presented in the form of a table occupying about one half of a page. Most textbooks presented arguments to emphasize the role of predictions; for instance: "It was the extraordinary success of Mendeléeff's predictions that led chemists not only to accept the periodic table but to recognize Mendeléeff more than anyone else as the originator of the concept on which it was based" (Bodner and Pardue, 1989, p. 201). According to Hill and Petrucci (1999): "The *predictive* nature of Mendeleev's periodic table led to its wide acceptance as tremendous scientific accomplishment" (p. 45). One of the textbooks (Phillips, Strozak, and Wistrom, 2000) compared the prediction of the elements and their properties to that of Halley's comet, which repeats its cycle every 76 years, and included an exercise in which the students are asked to predict the properties of an unknown element (Ge), while having the properties of Si, Ga, As, and Sn. Twenty-five textbooks reproduced Mendeleev's 1871 periodic table (at times in color and various devices to highlight missing elements), occupying about one half of a page to emphasize the elements predicted.

Relative Importance of Accommodation and Prediction in Development of the Periodic Table

None of the textbooks explained satisfactorily and only six made a simple mention of alternative interpretations with respect to the success of the periodic table (Brito et al 2005). One textbook came quite close to having a satisfactory presentation:

> Any good hypothesis must do two things: It must explain known facts, and it must make predictions about phenomena yet unknown . . . Mendeleev's hypothesis about how known chemical information could be organized passed all tests. Not only did the periodic table arrange data in a useful and consistent way to explain known facts about chemical reactivity, it also led to several remarkable predictions that were later found to be accurate (McMurry and Fay, 2001, p. 160).

This is a fairly good presentation of Mendeleev's dilemma (hypothesis) and could have been classified satisfactory had the authors recognized the role of controversy and alternative interpretations. The inclusion of alternative interpretations could facilitate students' understanding of how progress in science inevitably leads to controversies and rivalries and, at times, it is difficult to foresee and predict all implications of a theory.

Explanation of Periodicity in the Periodic Table

The objective behind this criterion was to make students think and reason with respect to the possible causes of periodicity in the periodic table. Many students must have wondered how a simple arrangement could provide such regularities. Textbooks could promote students' curiosity, and an historical reconstruction of the periodic table provides an opportunity to facilitate this objective by emphasizing:

a) inductive generalization, and
b) periodicity as a function of atomic theory. Apparently, none of the textbooks accomplished this objective satisfactorily, 14 made a simple mention and 43 simply ignored the issue (Brito et al. 2005). It is important to note that even those textbooks that ignored the issue implicitly recognized that the periodic table was a consequence of the accumulation of experimental data. Of the 14 textbooks that made a simple mention, some emphasized inductive generalization, and following are two examples:

Mendeleev's approach to the periodic table was empirical; he based his classification scheme on the observed facts. (Hill and Petrucci, 1999, p. 316)
The periodic table was created by Mendeleev to summarize experimental observations. He had no theory or model to explain why all alkaline earths combine with oxygen in a 1:1 atom ratio—they just do (Moore et al., 2002, p. 266).

In light of the historical reconstruction presented, to state that the periodic table was empirical and that Mendeleev had no theory or model to explain the periodicity of the properties of the elements is perhaps rather simplistic and difficult to sustain. It is more fruitful and plausible to present a more balanced picture to the students by highlighting the dilemma faced by Mendeleev (and others) in which they endeavored to look for underlying patterns to explain and understand periodicity.

Mendeleev's Contribution: Theory or Empirical Law?

This criterion is essential in understanding the nature of Mendeleev's contribution, viz., what exactly was he trying to do with all the information available. Mendeleev's own ambivalence notwithstanding, the historical reconstruction shows that Mendeleev's ingenuity consisted of precisely not only recognizing that the periodic table was a 'legitimate induction from the verified facts' but that there was a reason/cause/explanation for this periodicity, viz., the atomic theory. In other words, scientists do not decide beforehand that their contribution would be empirical / theoretical, but rather the scientific endeavor inevitably leads them to 'speculate' with respect to underlying patterns of what they observe. Mendeleev's case is an eloquent example of this dilemma. None of the textbooks made a satisfactory presentation and 52 simply ignored the issue (Brito et al 2005). Five textbooks made a simple mention and, of these, Lippincott et al. (1977) considered the periodic table to be an ordered domain: "If we examine the nature of scientific studies, we find that they always start with a group of observations collected as the data available for contemplation. The second step in the study is that of classification of data into recognizable related groupings…The periodic table …is an example of descriptive classification and ordering" (pp. 304–305). Stoker (1990) considered Mendeleev's contribution to be an empirical law:

> For many years after the formulation of the periodic law and the periodic table, both were considered to be empirical. The law worked and the table was very useful, but there was no explanation available for the law or for why the periodic table had the shape it had. It is now known that the theoretical basis for both the periodic law and the periodic table lies in electronic theory (p. 155).

This presentation is quite representative of most textbooks. It ignores the fact that scientists were constantly trying to look for a 'theoretical basis' of the periodic table, including Mendeleev himself. However, to state that for many years the table had no explanation is to ignore that progress in science is always tentative. In other words, our theories can hardly be considered to be final—in the future we may find a better explanation of the periodic table than that provided by the electronic theory. McMurry and Fay (2001) provide an example of how Mendeleev's contribution can be considered a theory:

> In many ways, the creation of the periodic table by Dmitri Mendeleev in 1869 is an ideal example of how a scientific theory comes into being. At first, there is only random information—a large number of elements and many observations about their properties and behavior. As more and more facts become known, people try to organize the data in ways that make sense, until ultimately a consistent hypothesis emerges (p. 160).

As we have observed in this study the development of the periodic table is much more complex. Nevertheless, recognition of the role played by 'emerging hypotheses' can facilitate a better understanding of the vicissitudes faced by Mendeleev and others, in their struggle to go beyond the observable entities.

Finally, textbooks give the impression that for almost 100 years (1820–1920) scientists had no idea or never asked the question as to whether there could be an underlying pattern to explain periodicity. In other words, textbooks provide students a non-controversial 'finished product' that could explain periodicity and the nature of the periodic table only when the modern atomic theory was formulated. The textbook approach does not facilitate students' understanding with respect to the tentative nature of science, considered to be important by modern philosophers of science and also science educators. Furthermore, neither science nor scientists can provide the final/true explanation.

Chapter 14

QUANTUM NUMBERS

Quantum numbers and electron configurations of chemical elements form an important part of the general chemistry curriculum and textbooks devote considerable amount of space to these topics. These topics are closely related to students' understanding of quantum mechanics and various studies have reported students' difficulties in grasping the fundamental issues involved (Hadzidaki et al. 2000; Kalkanis et al. 2003; Pospiech, 2000; Shiland, 1995, 1997; Taber 2005; Tsaparlis, 1997). Feynman (1985) was quite categorical: "… I can safely say that nobody understands quantum mechanics" (p. 129). On the other hand, philosophers of science have argued that quantum mechanics is particularly difficult to understand, due to the controversial nature of the different interpretations (Bohr's Copenhagen 'indeterminacy' and Bohm's 'hidden-variables' being two examples). In a recent critical review a physicist has conceded that:

> At the turn of the century, it is probably fair to say that we are no longer sure that the Copenhagen interpretation is the only possible consistent attitude for physicists … Alternative points of view are considered as perfectly consistent: theories including additional variables (or 'hidden variables') (Laloë 2001, p. 656).

Cushing (1991) has expressed the crux of the issue in cogent terms:

> The question is whether we are capable of truly *understanding* (or comprehending) quantum phenomena, as opposed to simply *accepting* the formalism and certain irreducible quantum correlations. The central issue is that

of understanding versus merely redefining terms to paper over our ignorance (p. 337, original italics).

In order to facilitate understanding, Cushing (1991) has suggested that in the physical sciences scientific theories function at the following levels: a) *Empirical adequacy*: Consists of essentially in 'getting the numbers right', in the sense of having a formula or an algorithm that is capable of reproducing observed data; b) *Explanation*: This is provided by a successful formalism with a set of equations and rules for its application; and c) *Understanding*: This is possible once we have an interpretation of the formalism that allows us to comprehend and to know the character of the phenomena and of the explanation offered.

A HISTORY AND PHILOSOPHY OF SCIENCE FRAMEWORK

Origin of the Quantum Hypothesis

It is well known that Thomas Kuhn directed the Project, 'Sources for History of Quantum Physics', a valuable archive now available at various institutions around the world. Besides a major publication (Heilbron and Kuhn, 1969), the only other scholarly work that was based on the sources of this 'Project' was about *Black-body theory* (Kuhn, 1978), which raised a provocative question: Who first proposed the quantum hypothesis? and stated categorically:

> the arguments in Planck's first quantum papers did not, as I now read them, seem to place any restrictions on the energy of the hypothetical resonators that their author had introduced to equilibrate the distribution of energy in the black-body radiation field. Planck's resonators, I concluded, absorbed and emitted energy continuously at a rate governed precisely by Maxwell's equations. His theory was still classical ... (p. viii)

Kuhn concluded that it was Paul Ehrenfest and Albert Einstein who first recognized that the black-body law could not be derived without restricting resonator energy to integral multiples of $h\nu$. In other words, Planck in 1900 simply introduced an approximate mathematical quantization for convenience in doing the calculations. On the other hand, the physical significance of the quantum hypothesis was first explained by Einstein. Despite skepticism on the part of some historians with respect to Kuhn's controversial thesis, Brush (2000) has shown that a historical reconstruction provides: " ... a clear confirmation of Kuhn's thesis" (p. 52).

In the context of Cushing's (1991) scheme, it is plausible to suggest that Planck's contribution in 1900 looks more like at the level of empirical adequacy, whereas Einstein's contribution provided some degree of explanation / understanding.

Alternative Interpretations of Quantum Mechanics

The standard or the orthodox view of quantum mechanics based on the Copenhagen interpretation (Bohr, Heisenberg, Pauli, Born) is almost universally accepted by chemists, physicists, philosophers of science and textbook authors. This interpretation requires complementarity (wave-particle duality), indeterminism, nonrealism and the impossibility of an event-by-event causal representation in a continuous space-time background. A review of the literature shows that both physicists and philosophers of science are increasingly getting convinced with respect to alternative interpretations (Bohm, 1952) of quantum mechanics. Cushing has invoked underdetermination of scientific theories by experimental evidence (Duhem-Quine thesis), to argue for the importance of alternative theories:

> In a sense, Bohm's 1952 work can be seen as an exercise in logic --- proving that Copenhagen dogma was not the only logical possibility compatible with the facts. In essence, Bohm accepted the formalism of quantum mechanics and showed that more microstructure is consistent with it than had previously been appreciated. By means of a reversible mathematical transformation, he was able to rewrite the Schrödinger equation in the form of Newton's second law of motion ('F = m a'). This result is exact and no approximations are made in obtaining it. In this theory, which predicts *all* of the results of standard quantum mechanics, there are event-by-event causality, a definite micro-ontology of actually existing particles that follow well defined trajectories in a space-time background --- just the type of thing that was *in principle* forbidden by the Copenhagen interpretation! (Cushing, 1995, p. 139).

This may sound strange to many chemistry teachers! The influence of the founding fathers of quantum mechanics (Bohr, Pauli, Heisenberg, Born) facilitated not only the establishment of the Copenhagen hegemony but also convinced the generation of physicists trained in the Copenhagen tradition to ignore Bohm's theory (Bohm, 1957, 1980; Olwell, 1999).

Differentiation between an Orbital and Electron Density

The common approach to the teaching of quantum numbers is based on the solutions of the Schrödinger equation for the one-electron atom. A review of the literature in chemistry education shows that a lack of an understanding with respect to quantum mechanics has led teachers and textbook authors to consider orbitals as physically observable rather than mathematical constructs. According to Ogilvie (1990, p. 285)

> orbitals *have no physical significance;* they are merely mathematical functions according to one particular approach (i.e., namely wave mechanics [Schrödinger], within its coordinate representation) to the mathematical solution ... In other words, there are no such *things* as orbitals.

The tetrahedral structure of methane has also been the subject of debate between Ogilvie (1990) and Pauling (1992). According to Ogilvie hybridization cannot explain the tetrahedral structure of methane. Pauling claimed that hybridization and the tetrahedral structure are verified experimentally (photoelectron spectrum) and that Ogilvie's arguments are illogical. This shows the difficulties involved in understanding quantum mechanics and its application to quantum numbers, electron configurations and related concepts such as hybridization.

Most general chemistry textbooks present quantum mechanics as a set of rules to allocate quantum numbers, which are later used to write electron configurations, based on further rules/algorithms, such as Pauli's exclusion principle, Aufbau principle and Hund's rule. There is almost no differentiation between an orbital (based on quantum mechanics) and electron density based on experimental measurements. Based on these considerations it is plausible to suggest that such presentations in Cushing's (1991) scheme can be classified at the level of 'empirical adequacy', which basically consists of 'getting the numbers right.' Gillespie et al. (1996) have suggested an alternative based on experimental measurements of ionization energies, which can provide an explanation and thus facilitate students' understanding. Richman (1998) has critiqued this approach on the grounds that it represents an empiricist epistemology that considers scientific theories to be primarily based on experimental data, viz., inductive generalizations. Niaz (1999) has argued that 'putting observations first' is not always the best teaching strategy. Nevertheless, in the case of quantum mechanics the complexity of the theoretical formulation is such that experimental determinations of electron density and ionizations energies can help to facilitate students' understanding.

Differentiation and Comparison between Classical and Quantum Mechanics

Teaching quantum mechanics both at the secondary and freshman university level becomes all the more difficult as students have already been exposed to classical Newtonian mechanics. Students' difficulties (alternative conceptions) in learning classical mechanics have been well documented in the literature (McCloskey et al. 1980). With respect to quantum mechanics, one study reported that freshman students:

> could not find a single reason that could relate 'wave-like' properties of particles to the behavior of electrons in the atom" (Kalkanis et al. 2003, p. 266).

Thus students have to not only overcome their alternative conceptions in classical mechanics but also learn the world of quantum phenomena. The same study also reported that:

> Bohr's 'planetary model' appeared to be deeply anchored in their cognitive systems: It was the atom model mainly remembered from secondary school chemistry classes ... this model was generally considered as the 'correct' one ... (Kalkanis et al. 2003, p. 266).

It is plausible to suggest that for many students, 'orbits' from Bohr's old quantum theory are perhaps the same as 'orbitals' in quantum mechanics (Taber, 2005).

To differentiate and compare classical and quantum mechanics, let us take the following example (Pospiech, 2000). Consider the photographs of a moving car: a) one with sharp contours and hence hiding the velocity; b) the other with blurred contours and hence hiding the exact position. In classical mechanics the car always has definite position and definite momentum. In quantum mechanics, however, the car would not be on a fixed path, only the photograph fixes the values, either for position or for momentum. The crucial point is that, statements can be made about the photograph only; the car itself is not accessible for further measurements.

Another example (Styer, 2000): One electron can be stripped away from a helium atom that is exposed to ultraviolet light below a certain wavelength. This threshold wavelength can be determined experimentally to very high accuracy. In contrast, classical mechanics predicts that any wavelength of light will strip away an electron.

To conceptualize such problems requires considerable cognitive effort on the part of the students. It is also quite apparent that any conceptual understanding of quantum mechanics will also require a reference to classical ways of thinking, or in other words quantum mechanics approaches classical mechanics as a limiting case:

> classical mechanics eventually gave way to the quantum theory, which is very different in its basic structure, but which still contains classical theory as a limiting case, valid approximately in the domain of large quantum numbers. Agreement with experiments in a limited domain and to a limited degree of approximation is evidently no proof, therefore, that the basic concepts of a given theory have a completely universal validity (Bohm, 1980, p. 82).

EVALUATION OF GENERAL CHEMISTRY TEXTBOOKS

Origin of the Quantum Hypothesis

Niaz and Fernández (in press) have reported that of the 55 textbooks analyzed, none described the origin of the quantum hypothesis satisfactorily. Russo and Silver (2002) was the only textbook that made a simple mention, in the following terms:

> when Bohr proposed that the energy of electrons in atoms is quantized, he was drawing on the work of the German physicist Max Planck. Planck had earlier proposed energy quantization in his efforts to explain the energy characteristics of the light emitted by heated objects. Planck saw this only as a mathematical trick, something that allowed him to arrive at the correct answer in his calculations. He did not believe that energy was actually quantized in any real physical system (p. 123).

This is an interesting presentation and could have been considered as satisfactory, if the authors would have mentioned how Einstein could be credited for the physical significance of quantization.

Most of the textbooks that made no mention, simply stated that Planck proposed and Einstein confirmed the quantum hypothesis, and following is an example:

> Planck proposed and Albert Einstein (1879-1955) confirmed that *the energy of a photon of electromagnetic radiation is proportional to its frequency,* not to its intensity or brightness as had been believed up to that time (Brady, et al. 2000, p. 281, original italics).

Alternative Interpretations of Quantum Mechanics

Four textbooks mentioned the role of alternative interpretations in quantum mechanics, and two described it satisfactorily (Niaz and Fernández, in press). Following is an example of a textbook that made a mention:

> Einstein, for instance, refused to believe in this newest quantum mechanical model and its probability-based view of the electron, insisting that 'God does not play dice with the universe.' Today, however, most physicists and chemists have been forced by the weight of many experimental observations to subscribe to Schrödinger's description of the atom, which is the current, modern, quantum mechanical model of the atom. Of course, some 'believers' have had to be dragged, kicking and screaming, to acceptance (Russo and Silver 2002, p.150, this is accompanied by a picture of a scientist kicking and screaming).

Such presentations by referring to Einstein's inconformity with quantum mechanics can arouse students' interest and curiosity. Nevertheless, these examples do not provide students with an alternative interpretation which could be pursued by students with a more inquiry oriented approach. Umland and Bellama (1999, p. 253) have innovated by going beyond and presenting the following section, and was classified as satisfactory: 'The Dirac equation: A relativistic model of the atom'. Although, this presentation does not explicitly refer to the underdetermination of scientific theories by experimental evidence it comes quite close to presenting an alternative interpretation and hence was classified as satisfactory.

Differentiation between an Orbital and Electron Density

None of the textbooks described satisfactorily that orbitals are mathematical constructs and the shapes of the orbitals (*s, p, d, f*) pictured in most textbooks are not derived from quantum mechanics but instead from electron density measurements. Following example, that was classified as no mention is quite representative of the dilemma faced by most textbooks:

> Wavefunctions of electrons in atoms are called atomic orbitals The mathematical expressions for atomic orbitals --- which are obtained as solutions of the Schrödinger equation are more complicated than the sine functions for the particle in a box ... the *square* of a wave function tells us the probability density of an electron at each point. To visualize this probability density, imagine a cloud centered on the nucleus. The density of the cloud at each point represents the probability of finding an electron there (Atkins and Jones 2002, p. 25).

Despite some positive aspects (atomic orbital being wavefunctions and obtained as solutions of the Schrödinger equation), this presentation makes no mention of the fact that representations of electron cloud densities are obtained from experimental measurements. Four textbooks made a simple mention and following is an example:

> The same theory tells us of the energies of atomic orbitals also describes their shapes. *How do we know these shapes are correct?* We don't for sure, but many of the predictions that have been made using the theory seem to be borne out by experiments. This gives the theoretical explanations strength and support. For example, the fact that the results of wave mechanics account for the shape of the periodic table so very well gives the theory a good deal of credibility (Brady, et al. 2000, p. 303, emphasis added).

Such presentations make an effort towards elucidating the shape of the orbitals by making students think as to how we know about the shape of the orbitals and how we can visualize them. Nevertheless, these textbooks fall short of being satisfactory as no explicit difference is established between orbitals as mathematical constructs and shape of the orbitals as a consequence of electron density measurements.

Differentiation and Comparison between Classical and Quantum Mechanics

As quantum mechanics approaches classical mechanics as a limiting case, and given students' difficulties with both, it is essential that textbooks facilitate conceptual understanding of quantum mechanics while having classical mechanics as a point of reference. In order to be classified as satisfactory, elaboration of an explicit framework based on teaching strategies / analogies that can facilitate understanding from classical to quantum mechanics is essential. Textbooks that did not elaborate any framework but provided at least two examples, illustrations, experiments or experiences to facilitate the transition were classified as mention. None of the textbooks analyzed were classified as satisfactory, whereas 28 textbooks made a simple mention (Niaz and Fernández, in press) and following are some of the examples:

> To understand all the properties of electromagnetic radiation, *both* a wave model and a particle model must be used. Some things are better explained by a wave model, and others are better explained by a particle model ... There is no analogy for the twofold character of electromagnetic radiation in everyday life. However, the picture in Figure 7.20 may help you to understand that the way something looks can depend on the way you look at it. Sometimes, when you look at Figure 7.20, you will see the young lady; other times you will see the old woman (Umland and Bellama 1999, p. 237)

> *Discussion:*
> What makes this presentation interesting is that the Figure 7.20 is reproduced from Hanson (1969, p. 90). Textbook does not acknowledge the source of the figure. Philosopher-physicist Norwood Russell Hanson used this figure to illustrate his thesis of theory-ladenness of observations, viz., what we observe is influenced by our prior theoretical frameworks.

> Quantization is like pouring water into a bucket. Water seems to be a continuous fluid, and it seems that any amount can be transferred. However, the smallest amount of water that we can transfer is one H_2O molecule. Likewise, energy seems to be continuously variable, but, in fact, it can be transferred only in discrete amounts (Jones and Atkins 2000, p. 272).

It is interesting to note that although none of the textbooks presented a framework to facilitate transition from classical to quantum mechanics, fair amount of attention is paid to including at least one or two strategies / analogies. The most common approaches seem to be that of a person climbing on a ladder and comparison of the wavelength of an electron and a baseball. Some of the textbooks went beyond and innovated by including thought provoking ideas, such as: picture of young/old woman, pouring water into a bucket, playing violin/piano, picture of a car with a high/low shutter speed, spokes of a bicycle wheel at low/high speed, and a fast-moving ballerina. The big challenge for textbook authors is to integrate these and other strategies/analogies into a framework that would be convincing and helpful to teachers. Despite such difficulties, it seems that textbooks generally do not go beyond what Cushing (1991) has referred to as the level of empirical adequacy.

Chapter 15

Conclusion

A historical reconstruction of the different topics of physical science shows that although experiments are important, interpretation of the data is even more important. In order to develop their research programs scientists rely on their guiding assumptions, which inevitably leads to conflicts and controversies. The rivalry between contending parties some times may last for many years and ultimately it is the scientific community that decides in favor of a particular theory. Even when textbooks present historical details it invariably is in the form of names of famous scientists including their pictures, year and place of work and anecdotes. Such presentations lack a framework based on history and philosophy of science and hence provide little additional insight to students as to how scientists work and theories are developed. Most students think that scientists are geniuses and they know what they are going to discover as soon as they enter the laboratory. This is indeed a caricature of what science is all about. It is concluded that if we want our students to understand and really scrutinize scientific practice, then a revision of the textbooks is necessary.

REFERENCES

Abd-El-Khalick, F., and Lederman, N.G. (2000). Improving science teachers' conceptions of nature of science: a critical review of the literature. *International Journal of Science Education, 22,* 665-701.

Achinstein, P. (1987). Scientific discovery and Maxwell's kinetic theory. *Philosophy of Science, 54,* 409–434.

Achinstein, P. (1991). *Particles and waves: Historical essays in the philosophy of science.* New York: Oxford University Press.

Akerson, V.L., Morrison, J.A., and McDuffie, A.R. (2006). One course is not enough: Preservice elementary teachers' retention of improved views of nature of science. *Journal of Research in Science Teaching, 43,* 194-213.

Allchin, D. (2005). Review of testing hypotheses: Prediction and prejudice. *Science, 308,* 1409-1410.

Alters, B.J. (1997). Whose nature of science? *Journal of Research in Science Teaching 34,* 39-55.

Andrews, T. (1869). On the continuity of the gaseous and liquid states of matter. *Scientific Papers,* (pp. 296–317, London, 1896).

American Association for the Advancement of Science, AAAS (1993) *Bench marks for science literacy: Project 2061.* New York: Oxford University Press.

American Association of Physics Teachers, AAPT (1999). What is science? *American Journal of Physics, 67,* 659.

Atkins, P. (1995). *The periodic kingdom. A journey into the land of the chemical elements.* New York: Harper Collins.

Atkins, P., and Jones, L. (2002). *Chemical Principles: The quest for insight* (2nd ed.). New York: Freeman.

Auerbach, D. (2000). What is science – isn't there more to it? *American Journal of Physics, 67,* 305.

Bell, R., Abd-El-Khalick, F., Lederman, N.G., McComas, W.F., and Matthews, M.R. (2001). The nature of science and science education: A bibliography. *Science and Education, 10*, 187-204.

Bensaude-Vincent, B. (1986). Mendeleev's periodic system of chemical elements. *British Journal for the History of Science, 19*, 3–17.

Blanco, R., and Niaz, M. (1997). Epistemological beliefs of students and teachers about the nature of science: From 'Baconian inductive ascent' to the 'irrelevance' of scientific laws. *Instructional Science, 25*, 203-231.

Bodner, G.M., and Pardue, H.L. (1989). *Chemistry: An experimental science.* New York: Wiley.

Bohm, D. (1952). A suggested interpretation of the quantum theory in terms of 'hidden' variables, I and II. *Physical Review, 85*, 166-179 and 180-193.

Bohm, D. (1957). *Causality and Chance in Modern Physics.* London: Routledge and Kegan Paul.

Bohm, D. (1980). *Wholeness and the Implicate Order.* London: Routledge and Kegan Paul.

Bohr, N. (1913). On the constitution of atoms and molecules. *Philosophical Magazine, 26*, 1–25.

Brady, J. E., Russell, J. W., and Holum, J. R. (2000). *Chemistry: Matter and its changes.* New York: Wiley.

Brito, A., Rodríguez, M.A., and Niaz, M. (2005). A reconstruction of development of the periodic table based on history and philosophy of science and its implications for general chemistry textbooks. *Journal of Research in Science Teaching, 42*, 84-111.

Brown, T.L., and LeMay, H.E. (1988). *Chemistry: The central science* (7th ed., Spanish). Englewood Cliffs, NJ: Prentice Hall.

Brush, S.G. (1976). *The kind of motion we call heat: A history of the kinetic theory of gases in the 19th Century.* New York: North-Holland.

Brush, S. (1996). The reception of Mendeleev's periodic law in America and Britain. *Isis, 87*, 595–628.

Brush, S.G. (2000). Thomas Kuhn as a historian of science. *Science and Education, 9*, 39–58.

Brush, S.G. (2005). Review of testing hypotheses: prediction and prejudice. *Science, 308*, 1410.

Brush, S.G. (2007). Predictivism and the periodic table. *Studies in History and Philosophy of Science, 38*, 256-259.

Burbules, N.C. and Linn, M.C. (1991). Science education and philosophy of science: Congruence or Contradiction? *International Journal of Science Education, 13*, 227-241.

Burns, R.A. (1996). *Fundamentals of chemistry* (2nd ed., Spanish). Englewood Cliffs, NJ: Prentice Hall.

Campanario, J.M. (2002). The parallelism between scientists' and students' resistance to new scientific ideas. *International Journal of Science Education, 24*, 1095-1110.

Chang, R. (1998). *Chemistry* (6th ed., Spanish). New York: McGraw-Hill.

Clark, P. (1976). Atomism versus thermodynamics. In C. Howson (Ed.), *Method and appraisal in the physical sciences: The critical background to modern science, 1800–1905* (pp. 41–105). Cambridge, U.K.: Cambridge University Press.

Clausius, R. (1857). On the nature of the motion which we call heat. *Philosophical Magazine, 14*, 108–127.

Clough, M.P. (2006). Learners' responses to the demands of conceptual change: Considerations for effective nature of science instruction. *Science and Education, 15*, 463-494.

Cohen, R. S. (1976). *Physical science*. New York: Holt, Rinehart and Winston.

Cooper, L. N. (1970). *An introduction to the meaning and structure of physics* (short edition). New York: Harper and Row.

Crowther, J. G. (1910). On the scattering of homogeneous β-rays and the number of electrons in the atom. Proceedings of the Royal Society, Vol. lxxxiv, London, pp. 226–247.

Cushing, J.T. (1991). Quantum theory and explanatory discourse: Endgame for understanding. *Philosophy of Science, 58*, 337-358.

Cushing, J.T. (1995). Quantum mechanics, underdetermination and hermeneutics. *Science and Education, 4* (2), 137-146.

De Berg, K.C. (2003). The development of the theory of electrolytic dissociation: A case study of a scientific controversy and the changing nature of chemistry. *Science and Education, 12*, 397-419.

De Regt, H.W. (1996). Philosophy and the kinetic theory of gases. *British Journal for the Philosophy of Science, 47*, 31–62.

Dickerson, R.E., Gray, H.B., Darensbourg, M.Y., and Darensbourg, D.J. (1984). *Chemical principles* (4th ed.) Menlo Park, CA: Benjamin Cummings.

Duhem, P. (1962). *The aim and structure of physical theory*. New York: Atheneum.

Ehrenhaft, F. (1910). Uber die kleinsten messbaren elektrizitätsmengen. Zweite vorläufige mitteilung der methode zur bestimmung des elektrischen elementarquantums. *Anzeiger Akad. Wiss* (Vienna) *10*, 118–119.

Eisberg, R. M. (1973). *Fundamentals of modern physics*. New York: Wiley.

Elkana, Y. (1974). *The interaction between science and philosophy*. Atlantic Highlands, NJ: Humanities Press.

Feynman, R. (1985). *The strange theory of light and matter*. London: Penguin.

Falconer, I. (1987). Corpuscles, electrons, and cathode Rays: J.J. Thomson and the 'Discovery of the Electron'. *British Journal for the History of Science, 20*, 241-276.

Forinash, K. (2000).What is science? *American Journal of Physics, 68*, 787–788.

Garber, E., Brush, S.G., and Everitt, C.W.F. Eds., (1986). *Maxwell on Molecules and Gases*. Cambridge, MA: MIT Press.

Gavroglu, K. (1990). The reaction of the British physicists and chemists to van der Waals' early work and to the law of corresponding states. *Historical Studies in the Physical and Biological Sciences, 20*, 199–237.

Gavroglu, K. (2000). Controversies and the becoming of physical chemistry. In P. Machamer, M. Pera, and A. Baltas (Eds.), *Scientific controversies: Philosophical and historical perspectives* (pp. 177–198). New York: Oxford University Press.

Geiger, H., and Marsden, E. (1909). On a diffuse reflection of the alpha particles. Proceedings of the Royal Society, Vol. lxxxii, London.

Gillespie, R.J., Spencer, J.N., and Moog, R.S. (1996). Demystifying introductory chemistry. Part 1. Electron configurations from experiment. *Journal of Chemical Education, 73*, 617-622.

Hadzidaki, P., Kalkanis, G., and Stavrou, D. (2000). Quantum mechanics: A systemic component of the modern physics paradigm. *Physics Education, 35*, 386-392.

Hanson, N.R. (1958). *Patterns of discovery*. Cambridge, UK: Cambridge University Press.

Hanson, N.R. (1969). *Perception and discovery*. San Francisco, CA: Freeman, Cooper and Co.

Heilbron, J. L. (1981a). Rutherford–Bohr atom. *American Journal of Physics, 49*, 223–231.

Heilbron, J. L. (1981b). *Historical studies in the theory of atomic structure*. New York: Arno Press.

Heilbron, J. L., and Kuhn, T. (1969). The genesis of the Bohr atom. *Historical Studies in the Physical Sciences, 1*, 211–290.

Hill, J.W. and Petrucci, R.H. (1999). *General chemistry: An integrated approach* (2nd ed.) Upper Saddle River, NJ: Prentice-Hall.

Hodson, D. (1988). Toward a philosophically more valid science curriculum. *Science Education, 72*, 19-40.

Holton, G. (1978). Subelectrons, presuppositions and the Millikan-Ehrenhaft dispute. *Historical Studies in the Physical Sciences, 9,* 161-224.

Holton, G. (1981). Foreword to the symposium: History of the atom. *American Journal of Physics, 49,* 205.

Holton, G. (1986). *The advancement of science and its burdens.* Cambridge, UK: Cambridge University Press.

Holton, G. (1992). Ernst Mach and the fortunes of positivism in America. *Isis, 83,* 27–60.

Holton, G. (1993). *Science and anti-science.* Cambridge, MA: Harvard University Press.

Holton, G. (1996). Science education and the sense of self. In P. R. Gross, and N. Levitt, and M.W. Lewis (Eds.), *The flight from science and reason* (pp. 551-560). New York: Academy of Sciences.

Hulsizer, R.I., and Lazarus, D. (1977). *The world of physics.* Menlo Park, CA: Addison–Wesley.

Ihde, A.J. (1969). Foreword. In J.W. van Spronsen (Ed.), *The periodic system of chemical elements: A history of the first hundred years.* Amsterdam: Elsevier.

Irwin, A.R. (2000). Historical case studies: Teaching the nature of science in context. *Science Education, 84,* 5-26.

Jaynes, E.T. (1967). Foundations of probability theory and statistical mechanics. *Delaware Seminar in the Foundations of Physics, 1,* 76–101.

Jones, R.C. (1995). The Millikan oil-drop experiment: Making it worthwhile. *American Journal of Physics, 63,* 970–977.

Jones, L., and Atkins, P. (2000). *Chemistry: Molecules, matter, and change* (4th ed.). New York: Freeman.

Jones, E. R., and Childers, R. L. (1990). *Contemporary college physics.* Menlo Park, CA: Addison-Wesley.

Justi, R. and Gilbert, J. (1999). A cause of ahistorical science teaching: Use of hybrid models. *Science Education, 83,* 163-177.

Kalkanis, G., Hadzidaki, P., and Stavrou, D. (2003). An instructional model for a radical conceptual change towards quantum mechanics concepts. *Science Education, 87,* 257-280.

Kaufmann, W. (1897). Die magnetische ablenkbarkeit der kathodenstrahlen und ihre abhangigkeit vom entladungspotential. *Annalen de Physik und Chemie, 61,* 544.

Kesidou, S., and Roseman, J.E. (2002). How well do middle school science programs measure up? Findings from Project 2061's curriculum review. *Journal of Research in Science Teaching, 39,* 522-549.

Kitchener, R.F. (1993). Piaget's epistemic subject and science education: Epistemological versus psychological issues. *Science and Education, 2*, 137–148.

Kohler, R.E. (1971). The origin of Lewis's theory of the shared pair bond. *Historical Studies in the Physical Sciences, 3*, 343-376.

Kotz, J.C., and Purcell, K.F. (1991). *Chemistry and chemical reactivity* (2 ed.). Philadelphia: Saunders.

Kuhn, T.S. (1970). *The Structure of Scientific Revolutions* (2nd edn.). Chicago: University of Chicago Press.

Kuhn, T.S. (1978). *Black-body theory and the quantum discontinuity: 1894-1912*. New York: Oxford University Press.

Lakatos, I. (1970). Falsification and the methodology of scientific research programmes. In I. Lakatos and A. Musgrave (Eds.), *Criticism and the growth of knowledge* (pp. 91–195). Cambridge, UK: Cambridge University Press.

Lakatos, I. (1971). History of science and its rational reconstructions. In R. C. Buck and R. S. Cohen (Eds.), *Boston Studies in the Philosophy of Science* (pp. 91-136). Dordrecht, The Netherlands: D. Reidel.

Laloë, F. (2001). Do we really understand quantum mechanics? Strange correlations, paradoxes, and theorems. *American Journal of Physics, 69*, 655-701.

Laudan, R., Laudan, L., and Donovan, A. (1988). Testing theories of scientific change. In A. Donovan, L. Laudan., and R. Laudan (Eds.), *Scrutinizing science: empirical studies of scientific change* (pp. 3-44). Dordrecht, The Netherlands: Kluwer.

Lederman, N.G. (1992). Students' and teachers' conceptions about the nature of science: A review of the research. *Journal of Research in Science Teaching, 29*, 331-359.

Lederman, N.G., Abd-El-Khalick, F., Bell, R.L., and Schwartz, R.S. (2002). Views of nature of science questionnaire: Toward valid and meaningful assessment of learners' conceptions of nature of science. *Journal of Research in Science Teaching, 39*, 497-521.

Lewis, G.N. (1916). The atom and the molecule. *Journal of the American Chemical Society, 38*, 1448-1455.

Lewis, G.N. (1923). *Valence and the structure of atoms and molecules*. New York: Chemical Catalog Co.

Lippincott, W. T., Garrett, A.B., and Verhoek, F.H. (1977). *Chemistry: A study of matter* (3rd ed.). New York: Wiley.

Lipton, P. (1991). *Inference to the best explanation*. London: Routledge.

Lipton, P. (2005). Testing hypotheses: prediction and prejudice. *Science, 307*, 219-221.

Mahan, B., and Myers, R.J. (1987). *University chemistry* (4th ed., Spanish). Menlo Park, CA: Benjamin Cummings.

Maher, P. (1988). Prediction, accommodation and the logic of discovery. In A. Fine and J. Leplin (Eds.), *PSA 1988: Vol. 1*. East Lansing, MI: Philosophy of Science Association.

Marquit, E. (1978). Philosophy of physics in general physics courses. *American Journal of Physics, 46*, 784-789.

Matthews, M.R. (1994). *Science teaching: The role of history and philosophy of science*. New York: Routledge.

Maxwell, J.C. (1860). Illustrations of the dynamical theory of gases. *Philosophical Magazine, 19*, 19–32. (*Scientific Papers*, 1965, pp. 377–409, New York: Dover).

Maxwell, J.C. (1875). On the dynamical evidence of the molecular constitution of bodies. *Journal of the Chemical Society* 13: 493–508. (*Scientific Papers*, 1965, pp. 418–438).

Maxwell, J.C. (1965). *Scientific Papers*. New York: Dover.

McCloskey, M., Caramazza, A., and Green, B. (1980). Curvilinear motion in the absence of external forces: Naïve beliefs about the motion of objects. *Science, 210*, 1139-1141.

McComas, W.F., Almazroa, H., and Clough, M.P. (1998). The role and character of the nature of science in science education. *Science and Education, 7*, 511-532.

McMullin, E. (1985). Galilean idealization. *Studies in History and Philosophy of Science, 16*, 247–273.

McMurry, J., and Fay, R.C. (2001). *Chemistry* (3rd ed.) Upper Saddle River, NJ: Prentice Hall.

Mellado, V., Ruiz, C., Bermejo, M.L., and Jiménez, R. (2006). Contributions from the philosophy of science to the education of science teachers. *Science and Education, 15*, 419-445.

Mendeleev, D. (1869). Ueber die beziehungen der eigenschaften zu den atom gewichten der elemente. *Zeitschrift für Chemie, 12*, 405–406 (English trans. by C. Giunta).

Mendeleev, D. (1879). The periodic law of the chemical elements. *The Chemical News, 40*, No. 1042.

Mendeleev, D. (1889). The periodic law of the chemical elements (Faraday lecture). *Journal of the Chemical Society, 55*, 634–656.

Mendeleev, D. (1897). *The principles of chemistry* (2nd English ed., trans. of sixth Russian ed.). New York: American Home Library Co.

Millikan, R.A. (1910). A new modification of the cloud method of determining the elementary electrical charge and the most probable value of that charge. *Philosophical Magazine 19*, 209–228.

Millikan, R.A. (1913). On the elementary electrical charge and the Avogadro constant. *Physical Review 2*, 109–143.

Millikan, R.A. (1916). The existence of a subelectron? *Physical Review 8*, 595–625.

Millikan, R.A. (1947). *Electrons (+ and –), Protons, Photons, Neutrons, and Cosmic Rays* (2nd ed.). Chicago: University of Chicago Press.

Monk, M., and Osborne, J. (1997). Placing the history and philosophy of science on the curriculum: A model for the development of pedagogy. *Science Education, 81*, 405-424.

Moore, J.W., Stanitski, C.L., and Jurs, P.C. (2002). *Chemistry: The molecular science*. Orlando, FL: Harcourt College.

Mortimer, C.E. (1983). *Chemistry* (5th ed.). Belmont, CA: Wadsworth.

Moseley, H.G.J. (1913). High frequency spectra of the elements. *Philosophical Magazine, 26*, 1025–1034.

National Research Council, NRC (1996). *National science education standards*. Washington, D.C.: National Academy Press.

Niaz, M. (1993). If Piaget's epistemic subject is dead, shall we bury the scientific research methodology of idealization? *Journal of Research in Science Teaching, 30*, 809–812.

Niaz, M. (1994). Enhancing thinking skills: Domain specific/domain general strategies --- A dilemma for science education'. *Instructional Science, 22*, 413-422.

Niaz, M. (1998). From cathode rays to alpha particles to quantum of action: A rational reconstruction of structure of the atom and its implications for chemistry textbooks. *Science Education, 82*, 527–552.

Niaz, M. (1999). Should we put observations first? *Journal of Chemical Education, 76*, 734.

Niaz, M. (2000a). The oil drop experiment: A rational reconstruction of the Millikan-Ehrenhaft controversy and its implications for chemistry textbooks. *Journal of Research in Science Teaching, 37*, 480-508.

Niaz, M. (2000b). A rational reconstruction of the kinetic molecular theory of gases based on history and philosophy of science and its implications for chemistry textbooks. *Instructional Science, 28*, 23-50.

Niaz, M. (2001a). Understanding nature of science as progressive transitions in heuristic principles. *Science Education, 85*, 684-690.

Niaz, M. (2001b). How important are the laws of definite and multiple proportions in chemistry and teaching chemistry? --- A history and philosophy of science perspective. *Science and Education, 10*, 243-266.

Niaz, M. (2001c). A rational reconstruction of the origin of the covalent bond and its implications for general chemistry textbooks. *International Journal of Science Education, 23*, 623-641.

Niaz, M. (2005). An appraisal of the controversial nature of the oil drop experiment: Is closure possible? *British Journal for the Philosophy of Science, 56*, 681-702.

Niaz, M. (2007). Progressive transitions in chemistry teachers' understanding of nature of science based on historical controversies. *Science and Education, 16*, in press.

Niaz, M., and Fernández, R. (in press). Understanding quantum numbers in general chemistry textbooks. *International Journal of Science Education*.

Niaz, M., and Robinson, W.R. (1992). From 'algorithmic mode' to 'conceptual gestalt' in understanding the behavior of gases: An epistemological perspective. *Research in Science and Technological Education, 10*, 53–64.

Niaz, M. and Robinson, W.R. (1993). Teaching algorithmic problem solving or conceptual understanding: Role of developmental level, mental capacity, and cognitive style. *Journal of Science Education and Technology, 2*, 407–416.

Niaz, M., and Rodríguez, M.A.(2005). The oil drop experiment: Do physical chemistry textbooks refer to its controversial nature? *Science and Education, 14*, 43-57.

Ogilivie, J.F. (1990). The nature of the chemical bond --- 1990: There are no such things as orbitals! *Journal of Chemical Education, 67*, 280-289.

Ohanian, H. C. (1987). *Modern physics*. Englewood Cliffs, NJ: Prentice-Hall.

Olenick, R.P., Apostol, T.M., and Goodstein, D.L. (1985). *Beyond the mechanical universe: From electricity to modern physics*. New York: Cambridge University Press.

Olwell, R. (1999). Physical isolation and marginalization in physics: David Bohm's cold war exile. *Isis, 90*, 738-756.

Osborne, J., Collins, S., Ratcliffe, M., Millar, R., and Duschl, R. (2003). What 'ideas-about-science' should be taught in school science? A Delphi study of the expert community. *Journal of Research in Science Teaching, 40*, 692-720.

Ostwald, W. (1927). *Lebenslinien, 2*, 178–179.

Oxtoby, D.W., Nachtrieb, N.H., and Freeman, W.A. (1990). *Chemistry: Science of change*. Philadelphia: Saunders.

Pauli, W. (1925). Über den zusammenhang des abschlußes der elektronengruppen im atom mit der komplexstruktur der spektren. *Zeitschrift für Physik, 31,* 765-785.

Pauling, L.(1952). *College chemistry.* San Francisco: Freeman.

Pauling, L. (1964). *General chemistry* (3rd ed.). San Francisco: Freeman.

Pauling, L. (1992). The nature of the chemical bond --- 1992. *Journal of Chemical Education, 69,* 519-521.

Phillips, J.S., Strozak, V.S., and Wistrom, C. (2000). *Chemistry: Concepts and applications* (Spanish ed.). New York: McGraw-Hill.

Polanyi, M. (1964). *Personal Knowledge.* Chicago: University of Chicago Press.

Pomeroy, D. (1993). Implications of teachers' beliefs about the nature of science. *Science Education, 77,* 261-278.

Pospiech, G. (2000). Uncertainty and complementarity: the heart of quantum physics. *Physics Education, 35,* 393-399.

Quagliano, J.V. and Vallarino, L.M. (1969). *Chemistry* (3rd ed.) Englewood Cliffs, NJ: Prentice-Hall.

Reinmuth, O. (1932). Editor's Outlook. *Journal of Chemical Education, 9,* 1139–1140.

Richman, R.M. (1998). In defense of quantum numbers. *Journal of Chemical Education, 75,* 536.

Rocke, A. J. (1984). *Chemical atomism in the nineteenth century: From Dalton to Cannizaro.* Columbus, Ohio: Ohio State University Press.

Rodebush, W.H. (1928). The electron theory of valence. *Chemical Review, 5,* 509-531.

Rodríguez, M.A., and Niaz, M. (2002). How in spite of the rhetoric, history of chemistry has been ignored in presenting atomic structure in textbooks. *Science and Education, 11,* 423-441.

Rodríguez, M., and Niaz, M. (2004a). A reconstruction of structure of the atom and its implications for general physics textbooks. *Journal of Science Education and Technology, 13,* 409-424.

Rodríguez, M.A., and Niaz, M. (2004b). The oil drop experiment: An illustration of scientific research methodology and its implications for physics textbooks. *Instructional Science, 32,* 357-386.

Rodríguez, M.A., and Niaz, M. (2004c) La teoría cinético-molecular de los gases en libros de física: Una perspectiva basada en la historia y filosofía de la ciencia, *Revista de Educación en Ciencias, 5,* 68-72.

Rosenfeld, L. (1963). *Introduction to Bohr's: On the Constitution of atoms and molecules.* Copenhagen, Denmark.

Russo, S., and Silver, M. (2002). *Introductory Chemistry* (2nd ed.). San Francisco: Benjamin Cummings.

Rutherford, E. (1911). The scattering of alpha and beta particles by matter and the structure of the atom. *Philosophical Magazine, 21*, 669–688.

Scharmann, L.C., and Smith, M.U. (2001). Further thoughts on defining versus describing the nature of science: A response to Niaz. *Science Education, 85*, 691-693.

Schwab, J.J. (1962). *The teaching of science as enquiry.* Cambridge, MA: Harvard University Press.

Schwab, J.J. (1974). The concept of the structure of a discipline. In E.W. Eisner and E. Vallance (Eds.), *Conflicting Conceptions of Curriculum* (pp. 162-175). Berkeley, CA: McCutchan.

Segal, B.G. (1989). *Chemistry: Experiment and theory* (2nd ed.) New York: Wiley.

Shapere, D. (1977). Scientific theories and their domains. In F. Suppe (Ed.). *The structure of scientific theories* (2nd ed., pp. 518–565). Chicago: University of Illinois Press.

Shiland, T.W. (1995). What's the use of all this theory? The role of quantum mechanics in high school chemistry textbooks. *Journal of Chemical Education, 72*, 215-219.

Shiland, T.W. (1997). Quantum mechanics and conceptual change in high school chemistry textbooks. *Journal of Research in Science Teaching, 34*, 535-545.

Sienko, M.J., and Plane, R.A. (1971). *Chemistry* (4th ed.) New York: McGraw-Hill.

Sisler, H.H., Dresdner, R.D., and Mooney, W.T. (1980). *Chemistry: A systematic approach.* New York: Oxford University Press.

Smith, M.U., Lederman, N.G., Bell, R.L., McComas, W.F., and Clough, M.P. (1997). How great is the disagreement about the nature of science: A response to Alters. *Journal of Research in Science Teaching, 34*, 1101-1103.

Smith, M.U., and Scharmann, L.C. (1999). Defining versus describing the nature of science: A pragmatic analysis for classroom teachers and science educators. *Science Education, 83*, 493-509.

Solomon, J., Scott, L., and Duveen, J. (1996). Large scale exploration of pupils' understanding of the nature of science. *Science Education, 80*, 493-508.

Stoker, H.S. (1990). *Introduction to chemical principles* (3rd ed.) New York: Macmillan.

Styer, D.F. (2000). *The strange world of quantum mechanics.* Cambridge, UK: Cambridge University Press.

Suvorov, S.G. (1966). Einstein's philosophical views and their relation to his physical opinions. *Soviet Physics Ospekhi, 8,* 578.
Taber, K.S. (2005). Learning quanta: Barriers to stimulating transitions in student understanding of orbital ideas. *Science Education, 89,* 94-116.
Taylor, H. S. (1942). The atomic concept of matter. In H. S. Taylor and S. Glasstone (Eds.), *A Treatise on physical chemistry.* Princeton, NJ: D. van Nostrand.
Thomson, J.J. (1897). Cathode rays. *Philosophical Magazine, 44,* 293-316.
Thomson, J.J. (1898). *Philosophical Magazine, 46,* 528.
Thomson, J.J. (1907). *The corpuscular theory of matter.* London: Constable.
Thomson, T. (1825). *An attempt to establish the first principles of chemistry by experiment.* London: Colburn and Bentley.
Townsend, J.S. (1897). *Proceedings of the Cambridge Philosophical Society, 9,* 244.
Tsaparlis, G. (1997). Atomic orbitals, molecular orbitals and related concepts: Conceptual difficulties among chemistry students. *Research in Science Education, 27,* 271-287.
Umland, J. B., and Bellama, J. M. (1999). *General chemistry* (3rd ed.). Pacific Grove, CA: Brooks/Cole.
van der Waals, J.D. (1873). *Over de continuiteit van den gas en vloeistoftoestand.*
van Spronsen, J. (1969). *The periodic system of chemical elements. A history of the first hundred years.* Amsterdam: Elsevier.
Wartofsky, M.W. (1968). *Conceptual foundations of scientific thought: An introduction to the philosophy of science.* New York: Macmillan.
Wiechert, E. (1897). Ergebniss einer messung der geschwindigkeit der cathodenstrahlen. *Schriften der Physikalischokonomisch Gesellschaft zu Konigsberg, 38,* 3.
Wilson, D. (1983). *Rutherford: Simple genius.* Cambridge, MA: MIT Press.
Wilson, J. D. (1996). *College physics* (2nd Spanish ed.). Englewood Cliffs, NJ: Prentice-Hall.
Ziman, J. (1978). *Reliable knowledge. An exploration of the grounds for belief in science.* Cambridge, UK: Cambridge University Press.
Zumdahl, S.S. (1993). *Chemistry* (3rd ed.) Lexington, MA: Heath.

INDEX

A

absolute truth, 1
accommodation, 52, 58, 81
accuracy, 68
algorithm, 64
algorithmic mode, 37, 43, 83
alkaline, 60
alternative, 3, 5, 26, 50, 58, 59, 60, 65, 66, 67, 69
aluminum, 52
ambiguity, 53, 54, 56
ambivalence, 53, 54, 56, 61
American Association for the Advancement of Science, 75
American Association of Physics Teachers (AAPT), 1, 2, 9, 75
American Physical Society, 1, 9
Amsterdam, 79, 86
approximation, 32, 41, 68
arbitrary assumptions, 35
argument, 13, 21
Aristotelian, 16, 30
aspiration, 56
assessment, 80
assumptions, vii, 5, 19, 21, 22, 24, 29, 30, 31, 32, 33, 41, 42, 73
atomic orbitals, 50, 70
atomic positions, 35
atomic structure, vii, 3, 9, 24, 54, 56, 78, 84
atomic theory, 25, 26, 27, 36, 53, 54, 55, 56, 60, 61, 62
atomic weight, 52, 54, 57
atomism, 84
atoms, 6, 7, 10, 11, 12, 15, 19, 36, 41, 46, 47, 48, 49, 50, 52, 53, 54, 55, 68, 70, 76, 80, 84
attacks, 12
attention, 18, 71
attitudes, 57
attribution, 57
authority, 12, 31

B

battery, 22, 23
behavior, 6, 15, 26, 32, 39, 41, 42, 43, 54, 62, 67, 83
beliefs, 76, 81, 84
beta particles, 85
black-body, 64
blocks, 51
Bohr, 9, 14, 15, 16, 17, 18, 53, 58, 65, 68, 76, 78
Boltzmann, 32, 37, 43
bonding, 45, 49
bonds, 45, 46, 50, 55
boron, 52
Boyle, 31, 32, 37, 43
Britain, 76
building blocks, 51

C

cadmium, 55
California, 21
carbon dioxide, 33
case study, 77
catalyst, 53
cathode ray tube, 19
cathode rays, 6, 9, 10, 82
causality, 65
ceteris paribus clauses, 30, 43
chaos, 15
Charles, 32, 37
chemical, 35, 49, 51, 52, 53, 54, 55, 57, 58, 59, 63, 75, 76, 79, 80, 81, 83, 84, 85, 86
chemical properties, 52, 53
chemical reactivity, 59, 80
chemistry, vii, 3, 10, 13, 16, 17, 18, 19, 22, 23, 24, 25, 26, 41, 42, 43, 45, 48, 51, 53, 54, 58, 63, 65, 66, 67, 76, 77, 78, 81, 82, 83, 84, 85, 86
Chicago, 9, 19, 80, 82, 84, 85
chlorides, 52
classes, 67
classical electrodynamics, 14, 17
classical mechanics, 67, 68, 71
classification, 45, 51, 57, 60, 61
classroom teachers, 85
Clausius, 29, 30, 33, 37, 77
closure, 83
co-existence, 16
cognitive effort, 68
cognitive style, 83
cognitive system, 67
cohesion, 33
cold war, 83
collisions, 12, 13, 29, 32
community, 7, 15, 20, 31, 73, 83
competition, 7
complementarity, 65, 84
complexity, 1, 66
components, 5, 30
composition, 25
compounds, 26, 46, 47, 51
conception, 11, 52, 54, 56
conceptual gestalt, 37, 83
configuration, 49, 58
conflict, 12
consensus, 7, 45
construction, 5, 18, 31
continuity, 33, 75
convection, 22
conviction, 20
Copenhagen, 63, 65, 84
correlations, 63, 80
covalent bond, 45, 47, 48, 49, 50, 83
creativity, 53
credibility, 1, 70
criticism, 20, 35, 56
curiosity, 60, 69
curriculum, 2, 63, 78, 79, 82

D

data base, vii, 7
dating, 52
deduction, 32
defense, 20, 84
definition, 1, 2
degrees of freedom, 31
Delaware, 79
Delphi study, 83
Denmark, 84
density, 22, 23, 52, 57, 67, 70
differentiation, 66
Dirac equation, 69
directives, 31
discipline, 85
discontinuity, 80
discourse, 77
dissociation, 77
distribution, 13, 30, 31, 64
diversity, 43

E

education, vii, 2, 3, 5, 7, 18, 45, 51, 66, 75, 76, 77, 78, 79, 80, 81, 82, 83, 84, 85, 86
educators, 2, 62, 85

Einstein, Albert, 64, 68, 69
elaboration, 47, 71
electric charge, 22
electric field, 10, 11, 20
electricity, 6, 22, 83
electromagnetic, 16, 69, 71
electron(s), 7, 10, 11, 14, 15, 16, 17, 18, 19, 20, 21, 22, 45, 46, 47, 48, 49, 50, 63, 66, 67, 68, 69, 70, 72, 77, 78, 84
electron charge, 22
electron density, 66, 70
electrostatic force, 6
elementary electrical charge, vii, 3, 19, 20, 21, 22, 23, 82
elementary particle, 31
elementary teachers, 75
empirical data, vii, 5
energetics, 35
energy, 17, 18, 49, 64, 68, 69, 71
energy characteristics, 68
England, 10
epistemology, 5, 27, 66
ethene, 25
ether, 9
European, 20
evaporation, 22, 23
evidence, 1, 7, 12, 21, 27, 49, 50, 59, 65, 69, 81
exclusion, 47, 50, 66
exercise, 54, 59, 65
experimental condition, 22, 23
experimental data, 1, 6, 7, 11, 24, 37, 60, 66

F

Feynman, 63, 78
first principles, 86
flavor, 20
flight, 79
fluid, 71
fluorine, 48
freedom, 31
freshman, vii, 3, 24, 39, 45, 67
friction, 23

G

gallium, 52, 54
gases, vii, 3, 20, 29, 30, 32, 33, 37, 39, 42, 43, 49, 76, 77, 81, 82, 83, 84
gasoline, 49
Gay-Lusaac, 26, 27, 32, 37
Gay-Lusaac's empirical law, 26
gene, 26, 66
generalization(s), 26, 47, 49, 53, 60, 66
generation, 30, 65
germanium, 52
Gestalt, 39
God, 69
gold, 11, 13, 14
gravity, 23
growth, 16, 33, 80
guidelines, 11, 18

H

handwriting, 21
Harvard, 79, 85
heart, 43, 84
heat, 30, 52, 76, 77
hegemony, 65
Heisenberg, 65
helium, 68
heuristic, 1, 2, 5, 6, 7, 9, 10, 13, 16, 17, 18, 22, 23, 24, 26, 31, 32, 33, 41, 43, 48, 58, 83
heuristic principles, 5
hidden-variables, 63
high school, 85
historian, 2, 51, 53, 76
hybrid, 79
hybridization, 66
hydrocarbons, 25
hydrogen, 17, 18, 46, 49, 50
hydrogen atoms, 50
hypothesis, 11, 13, 14, 18, 31, 35, 52, 54, 57, 59, 60, 62, 64, 68, 69

Index

I

Ideal Gas Law, 37
idealization, 30, 37, 81, 82
imagination, 7
incidence, 24
inclusion, vii, 2, 3, 9, 60
indeterminacy, 63
indeterminism, 65
indication, 5
induction, 56, 61
Inductivist, 26, 27, 49
infinite, 20
insight, 20, 73, 75
instability, 14
institutions, 64
instruction, 77
intensity, 69
interaction(s), 31, 78
interpretation, vii, 15, 17, 26, 27, 36, 47, 50, 63, 64, 65, 69, 73, 76
interval, 29
intuition, 36
ionization, 66
ions, 6, 10, 20, 48
isolation, 83
isotherms, 33

J

Jaynes, 36, 79
judgment, 18
justification, 33

K

kernel, 48
Kinetic theory, vii

L

Lakatosian, 26, 27, 31, 49, 50
land, 75
law of multiple proportions, 25, 26, 53, 55, 56
laws, 1, 7, 14, 15, 16, 25, 26, 27, 31, 32, 37, 41, 43, 47, 54, 57, 76, 83
lead, 3, 17, 21, 26, 45, 47, 55
learners, 80
learning, 2, 67
liquid phase, 33
liquids, 29
literacy, 3, 75
literature, vii, 5, 21, 22, 25, 29, 39, 65, 66, 67, 75
London, 75, 76, 77, 78, 80, 86

M

manipulation, 37
marginalization, 83
Maxwell, 16, 17, 29, 30, 31, 32, 33, 35, 37, 41, 42, 43, 64, 75, 78, 81
Maxwell's theory, 33
meanings, 54
melting, 52
memory, 51
Mendeleev, 51, 52, 53, 54, 55, 56, 57, 58, 59, 60, 61, 62, 76, 81, 82
Mendeleev's hypothesis, 59
mental capacity, 83
methane, 25, 46, 66
microstructure, 65
MIT, 78, 86
model(s), 3, 9, 10, 11, 12, 13, 14, 15, 16, 17, 18, 25, 30, 35, 37, 41, 42, 45, 47, 49, 50, 60, 67, 69, 71, 79, 82
molecular forces, 29, 33
molecular model, 33
molecular orbitals, 86
molecular weight, 54
molecules, 6, 15, 27, 29, 31, 41, 47, 52, 53, 76, 80, 84
momentum, 31, 67
monotheistic Popperianism, 1
motion, 22, 30, 31, 65, 76, 77, 81
movement, 30
multiples, 22, 64

Index

N

Na$^+$, 46
National Research Council, 3, 82
Netherlands, 80
New York, 75, 76, 77, 78, 79, 80, 81, 82, 83, 84, 85, 86
Newton's law, 31
Newton's second law, 65
Newtonian, 15, 30, 32, 42, 67
noble gases, 49
NRC, 3, 82
nucleus(i), 12, 16, 18, 50, 70

O

observations, vii, 7, 11, 15, 18, 20, 21, 22, 26, 47, 54, 56, 60, 61, 62, 66, 69, 71, 82
observed behavior, 41
oil, 11, 13, 14, 19, 23, 24, 79, 82, 83, 84
oil drop experiment, 19, 23, 24, 82, 83, 84
organization, 51
Ostwald, W., 83
oxide(s), 22, 52, 55, 57
oxygen, 55, 60

P

Pacific, 86
parallelism, 77
particles, 6, 7, 9, 10, 11, 13, 22, 30, 31, 32, 42, 65, 67, 78, 82, 85
pattern, 22, 37, 62
Pauling, Linus, 25
pedagogical, 6
pedagogy, 82
peer review, 7
peers, 6
periodic law, 45, 51, 52, 54, 55, 56, 58, 59, 61, 76, 81
Periodic Table, vii, 3, 51, 53, 58, 59, 60
periodicity, 53, 54, 55, 56, 60, 61, 62
personal, 2
personality, 57
philosophers, 2, 5, 11, 17, 19, 35, 51, 52, 56, 62, 63, 65
philosophical, 5, 15, 18, 35, 57, 86
philosophy of science, vii, 1, 2, 25, 73
photographs, 67
photon, 69
physical chemistry, 24, 78, 83, 86
physical properties, 52
physical sciences, 6, 36, 64, 77
physicist, 2, 6, 10, 15, 63, 68, 71
physicochemical properties, 52, 58
physics, vii, 3, 5, 6, 10, 12, 13, 16, 17, 18, 19, 22, 24, 42, 43, 77, 78, 79, 81, 83, 84, 86
plausibility, 20, 30
positivism, 56, 79
positivist, vii, 5, 15, 17, 25, 35, 57
positivistic, 36
power, 1, 43, 57
prediction, 52, 57, 59, 76, 81
predictivist thesis, 52
prejudice(s), 36, 75, 76, 81
pressure, 22, 30, 32
prestige, 12
probability, 11, 50, 69, 70, 79
probability theory, 79
problem solving, 83
program, 9, 19, 30, 31, 32, 35, 42, 49
promote, 35, 60

Q

quanta, 15, 86
quantization, 17, 64, 68
quantum mechanics, 63, 65, 66, 67, 68, 69, 70, 71, 79, 80, 85
quantum numbers, vii, 3, 63
quantum phenomena, 63, 67
quantum theory, 47, 67, 68, 76
questionnaire, 80

R

radiation, 15, 64, 69, 71
radius, 20, 22, 23

range, 22, 41
rational reconstruction, 80, 82, 83
reactivity, 59, 80
reading, 12, 19, 30
reality, 46
reasoning, 6
recall, 53
reception, 2, 15, 76
recognition, 52, 62
reconcile, 20
reconstruction, 2, 9, 11, 15, 16, 19, 50, 53, 60, 61, 65, 73, 76, 82, 83, 84
reduction, 21
reflection, 57, 78
rejection, 55, 57
relationship(s), 20, 26, 27, 51, 54, 55
relevance, 51
resistance, 77
resolution, 37
resonator, 64
retention, 75
rhetoric, 10, 18, 84
Royal Society, 12, 77, 78
Rutherford, 9, 11, 12, 13, 14, 15, 16, 17, 18, 21, 53, 78, 85, 86

S

scandium, 52, 54
scattering, 11, 13, 77, 85
school, 35, 67, 79, 83, 85
Schrödinger equation, 65, 66, 70
science, vii, 1, 2, 3, 5, 6, 7, 11, 12, 15, 16, 17, 18, 19, 24, 25, 26, 27, 29, 30, 31, 35, 42, 44, 45, 46, 48, 51, 52, 56, 57, 58, 60, 61, 62, 63, 65, 73, 75, 76, 77, 78, 79, 80, 81, 82, 83, 84, 85, 86
science curricula, 5, 29
science educators, 2, 62, 85
science literacy, 3, 75
science teaching, 79
scientific community, 15, 20, 73
scientific knowledge, 1, 5, 50
scientific method, 7
scientific progress, 3, 15, 43, 46, 47, 58

scientific theory(ies), 1, 7, 26, 47, 57, 62, 64, 65, 66, 69, 85
scientists, vii, 1, 3, 7, 9, 15, 16, 17, 23, 26, 31, 35, 37, 43, 49, 52, 56, 61, 62, 73, 77
search, 22, 57
series, 6, 14, 15, 47
shape, 61, 70
sharing, 45, 47, 49, 50
silicon, 52, 57
silver, 55
simplifying assumptions, 30, 32, 33, 35, 41
sine, 70
skills, 82
space-time, 65
specific heat, 52
spectrum, 14, 15, 17, 66
speculation, 7, 56
speed, 72
spin, 6, 50
stability, 14, 16, 17, 49, 50
stages, 47
standards, 3, 82
statistical mechanics, 79
statistics, 31
strategies, 10, 71, 82
strength, 50, 70
students, vii, 2, 5, 6, 9, 23, 39, 45, 49, 51, 53, 58, 59, 60, 62, 63, 66, 67, 68, 69, 70, 71, 73, 76, 77, 86
students' understanding, vii, 39, 51, 58, 60, 62, 63, 66
sulfides, 52
surprise, 5
symbiosis, 16
systems, 14, 41, 54, 67

T

teachers, 1, 5, 11, 42, 45, 51, 65, 66, 72, 75, 76, 80, 81, 83, 84, 85
teaching, 2, 66, 71, 79, 81, 83, 85
teaching strategies, 71
temperature, 22, 23, 30, 33
textbooks, vii, 2, 3, 5, 6, 9, 10, 13, 16, 17, 18, 19, 22, 23, 24, 25, 26, 27, 29, 41, 42, 43,

45, 47, 48, 49, 50, 53, 58, 59, 60, 61, 62, 63, 66, 68, 69, 70, 71, 73, 76, 82, 83, 84, 85
theory, 2, 3, 7, 11, 16, 17, 18, 25, 26, 27, 29, 30, 31, 32, 33, 35, 37, 39, 41, 42, 43, 45, 46, 47, 53, 54, 55, 56, 57, 58, 60, 61, 62, 64, 65, 67, 68, 70, 71, 73, 75, 76, 77, 78, 79, 80, 81, 82, 84, 85, 86
thermodynamics, 29, 35, 77
thinking, 36, 68, 82
Thomson, 6, 9, 10, 11, 12, 13, 14, 18, 19, 24, 25, 33, 46, 53, 58, 78, 86
threshold, 68
time, 11, 12, 21, 29, 52, 65, 69
tin, 57
tradition, 57, 65
transformation, 65
transition(s), 2, 33, 43, 71, 83, 86
transport, 32

U

UK, 78, 79, 80, 85, 86
ultraviolet light, 68
uniform, 30, 55
universe, 69, 83

V

valence, 47, 48, 49, 52, 53, 54, 84
validity, 52, 57, 68
values, 19, 20, 32, 59, 67
van der Waals, 33, 42, 78, 86
variable(s), 23, 24, 37, 63, 71, 76
variation, 22
vein, 36
velocity, 23, 30, 67
viscosity, 23
voice, 2

W

war, 83
Washington, 82
writing, 45

X

X-ray, 19